Cellular System Design and Optimization

Clint Smith, P.E.

Curt Gervelis

McGraw-Hill

New York   San Francisco   Washington, D.C.   Auckland   Bogotá
Caracas   Lisbon   London   Madrid   Mexico City   Milan
Montreal   New Delhi   San Juan   Singapore
Sydney   Tokyo   Toronto

Library of Congress Cataloging-in-Publication Data

Smith, Clint
    Cellular system design and optimization / Clint Smith, Curt Gervelis.
      p.   cm.
    Includes index
    ISBN 0-07-059273-x
    1. Wireless communication systems.   2. Cellular Radio.
  I. Gervelis, Curt. II. Title
TK5103.2.S65   1996
621.3845′6—dc20
                                  96-14902
                                       CIP

McGraw-Hill

A Division of The McGraw·Hill Companies

1 2 3 4 5 6 7 8 9 0 DOC/DOC 9 0 1 0 9 8 7 6

ISBN 0-07-059273-X

The sponsoring editor for this book was Stephen S. Chapman, the editing supervisor was David E. Fogarty, and the production supervisor was Pamela A. Pelton. It was set in Century Schoolbook by Dina E. John of McGraw-Hill's Professional Book Group composition unit.

Printed and bound by R. R. Donnelley & Sons Company.

This book is printed on acid-free paper.

To Sam and Mary for their continued support and patience during this endeavor

To our caring parents

And to our many colleagues who have shared their time and inspiration throughout these past exciting years in the cellular industry

Contents

Preface

The intention behind writing this book has been to transfer real-life situations and solutions into a formal document for cellular engineers to utilize. We have strived to compile a large amount of information in a concise book whose target audience is the cellular engineering community. We have trained many engineers in the industry and have found a consistent lack of available written practical guidance. Through conveying what has and has not been successful in engineering we hope to avoid having many common mistakes continue to recur.

This book is a general guide or rather a book of knowledge to be used by cellular and wireless engineers, experienced and novice. The information contained here is intended for the people who have to make a system work on a daily basis. As our colleague George Starace once said, "You must always concentrate on where you are getting your nickels from."

This guide covers both the RF and network aspects of the engineering efforts associated with a cellular system. We begin by introducing cellular communication and cover basic RF and network concepts. Design guidelines are then covered with numerous examples of actual troubleshooting techniques. We then discuss how to monitor a system, put together a growth study, and organize engineering. The book's general flow is to discuss the key topics first and then go over examples and/or case studies that can be used as a general reference.

Chapter 1 is an introduction to cellular communication, presenting the general designs and concepts. The chapter includes an overview of a general cellular network, MTSO, and cell site configuration. Additional topics include call scenarios, handoff spectrum allocation, and frequency reuse.

Chapter 2 addresses topics basic to radio-frequency engineering. This chapter goes over basic radio, covering propagation, receivers, antennas, transmitters, filters, ERP, link budgets, and cell site configurations.

The focus of Chapter 3 is on basic network components and concepts. The intention here is to give a basic overall description of some of the major parts of the cellular network. Some aspects covered are basic network components, switching concepts, basic telephony, T-carriers, PCM, LEC, and IXC. Discussions also include SS7 concepts and networks.

Chapter 4 addresses RF engineering design guidelines. This chapter goes over the design guideline philosophies to utilize, design reviews, required signatures, design change orders, and tracking mechanisms. Topics included here focus on the cell site design phases involving selection of the search area, site acceptance or rejection, FAA and FCC rules, EMF, and some aspects of the zoning process. Other design criteria topics involve radio expansion guidelines, frequency planning, antenna selection and change, cell site parameter setting and adjustment, and software.

Network design guidelines make up Chapter 5. This chapter goes over the network design guideline philosophy to utilize for network designs. System interconnect design for voice and data are other topics included. The SS7 network basic design parameters and cell site data link design issues are also covered in this chapter.

System performance aspects of an RF and network are covered in Chapters 6 and 7. This is the area that tends to separate engineers into good, mediocre, and outright poor. These chapters focus on how you monitor and optimize the network on a continuous basis including recommended fixes that have worked, independent of the vendor. Also included are equations and suggested tips for expediently resolving performance issues with examples of what has worked, plus what has not worked, and attempts to try to explain why it occurred. Topics included in these chapters are lost calls, intermodulation, radio and trunk blocking, directed retry, access problems, downtilting, antenna orientation, drive testing, retunes, EMF, MOPs, design reviews, CPU load studies, data link performance studies, sample call delivery senarios, and network timing problems.

Chapter 8 addresses the much needed issue of what reports should be generated, how often, and who should be the recipient of each. A hierarchical approach to report generation is presented. For example, a report to an engineer should be different from the one a manager, director, or vice president uses. There are also different reports between cross-functional departments like operations engineering and construction that need to be conveyed. Examples of general reports that are covered include network performance reports, RF performance reports, exception reports, customer care reports, software configuration reports, and project reports (current and pending). Also addressed are when to conduct meetings, focusing on what types of meetings they should be, who should attend, and how frequent they should be.

The methods and procedures needed to put together a network growth plan are covered in Chapter 9. This chapter explains how to assemble a growth plan for a network that is in existence, specifically what should be included, where you get the information, and how to actually complete a study, with a few examples. The procedure shows how to determine the growth aspects for a network through identifying what key parameters are used when defining its growth. Some aspects of switch growth involve defining the current growth of a switch's ports, CPU loading, subscriber datebase limits, and disaster recovery. Cell site aspects of a growth plan are design criteria, portability, mobility, frequency planning, radio growth, physical equipment capacity, software loads, prioritization, and real estate acquisition.

Chapter 10 covers the area of organization and training for an engineering department. This chapter addresses the two issues of organization layout and technical training recommendations for various engineering levels. The chapter also presents a proposed schedule, pointing out the depth and breadth needed for each area or discipline. The intention here is to provide a general guideline of training for an engineering group at different levels. Lastly, a suggestion on some basic material for an engineering library is also included in order to establish basic reference sources.

In closing we both trust that the information in this book will further the cellular engineering profession as it continues to mature and develop.

Clint Smith, P.E.
Curt Gervelis

Cellular
System Design
and Optimization

1

Cellular Communication

This chapter briefly covers some of the key concepts of cellular communication. Its objective is to provide an overview of certain concepts discussed in subsequent chapters. The concept of cellular radio was initially developed by AT&T at their Bell Laboratories to provide additional radio capacity for a geographic customer service area. The initial mobile systems from which cellular radio evolved were called mobile telephone systems (MTSs). Later improvements to these systems occurred, and the systems were referred to as improved mobile telephone systems (IMTSs). One of the main problems with these systems was that a mobile call could not be transferred from one radio station to another without loss of communication. This problem was relieved by reusing the allocated frequencies of the system, enabling a market to offer higher radio traffic capacity. The increased radio traffic allows more users in a geographic service area than with the MTS or IMTS. A list of current cellular system types is provided below.

Cellular technologies

Advanced mobile phone system (AMPS) standard. AMPS is the cellular standard that was developed for use in North America. This type of system operates in the 800-MHz frequency band. AMPS systems have also been deployed in South America, Asia, and Russia.

Code-division multiple access (CDMA). CDMA is an alternative digital cellular standard developed in the United States. No systems of this type are in operation at this time.

Digital AMPS (D-AMPS). D-AMPS [also known as time-division multiple access (TDMA) and/or IS-54] is the digital standard for cellular systems developed for use in the United States. The AMPS standard was developed into the D-AMPS digital standard instead of a completely new standard. This was done to quickly provide a means to expand the existing analog systems that were growing at a rapid pace.

DCS1800. DCS1800 is a digital standard based upon the GSM technology, with the exception that this type of system operates at a higher frequency range, 1800 MHz. The DCS1800 technology is intended for use in PCS-type systems. Systems of this type have been installed in Germany and England.

Global System for Mobile communications (GSM). GSM is the European standard for digital cellular systems operating in the 900-MHz band. This technology was developed out of the need for increased service capacity due to the analog system's limited growth. This technology offers international roaming, high speech quality, increased security, and the ability to develop advanced systems features. The development of this technology was completed by a consortium of 80 pan-European countries working together to provide integrated cellular systems across different borders and cultures.

Nordic Mobile Telephone (NMT) standard. NMT is the cellular standard that was developed by the Nordic countries of Sweden, Denmark, Finland, and Normandy in 1981. This type of system was designed to operate in the 450- and 900-MHz frequency bands. These are noted as NMT 450 and NMT 900. NMT systems have also been deployed throughout Europe, Asia, and Australia.

Personal communication system (PCS). PCS is a general name given to systems that have recently developed out of the need for more capacity and design flexibility than that provided by the initial cellular systems.

Personal Digital Cellular (PDC) standard. PDC is a digital cellular standard developed by Japan. PDC-type systems were designed to operate in the 800-MHz and 1.5-GHz bands.

Total Access Communications Systems (TACS). TACS is a cellular standard that was derived from the AMPS technology. TACS systems operate in both the 800- and 900-MHz band. The first system of this kind was implemented in England. Later these systems were installed in Europe, China, Hong Kong, Singapore, and the Middle East. A variation of this standard was implemented in Japan.

1.1   The Cellular Concept

Cellular radio was a logical progression in the quest to provide additional radio capacity for a geographic area. The cellular system as it is known today has its primary roots in the MTS and IMTS, which are similar to cellular except that no handoff takes place with these networks.

Cellular systems operate on the principle of frequency reuse. Frequency reuse in a cellular market gives a cellular operator the ability to offer higher radio traffic capacity. The higher radio traffic capacity enables many more users in a geographic area to utilize radio communication than are available with an MTS or IMTS system.

The cellular systems are broken into metropolitan statistical areas (MSAs), and rural statistical areas (RSAs). Each MSA and RSA has two different cellular operators that offer service. They are referred to as A-band and B-band systems. The A-band system is the nonwireline system and the B-band is the wireline system for the MSA or RSA.

1.2   Generic Cellular System Configuration

A generic cellular system configuration is shown in Fig. 1.1. It involves all the high-level system blocks of a cellular network. Many components comprise each of the blocks shown. Each of the individual system components of a cellular network is covered in later chapters of this book. Referring to Fig. 1.1, the mobile communicates to the cell site through use of radio transmission, which utilize a full duplex configuration that involves a separate transmit and receive frequency used by the mobile and cell site. The cell site transmits on the frequency the mobile unit is tuned to, and the mobile unit transmits on the radio frequency the cell site receiver is tuned to.

The cell site acts as a conduit for the information transfer converting the radio energy into another medium. The cell site sends and receives information from the mobile and the mobile telephone system office (MTSO). The MTSO is connected to the cell site either by

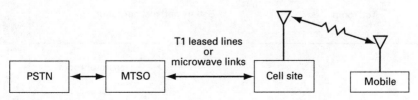

Figure 1.1   Simplified cellular system.

leased T1 lines or through a microwave system. The cellular system is made up of many cell sites which all interconnect back to the MTSO.

The MTSO processes the call and connects the cell site radio link to the public switched telephone network (PSTN). The MTSO performs a variety of functions involved with call processing and is effectively the brains of the network. The MTSO maintains the individual subscriber records, the current status of the subscribers, call routing, and billing information, to mention a few items.

1.3   Generic MTSO Configuration

Figure 1.2 is a generic mobile telephone switching office configuration. The MTSO is the portion of the network which interfaces the radio world to the public service telephone network. In mature systems there are often multiple MTSO locations, and each MTSO can have several cellular switches located within each building.

1.4   Generic Cell Site Configuration

Figure 1.3 is an example of a generic cell site configuration. The cell site configuration shown in Fig. 1.3 is a picture of a monopole cell site. An equipment hut associated with it houses the radio-transmission equipment. The monopole, which is next to the equipment hut, supports the antennas used for the cell site at the very top of the monopole. The cable tray between the equipment hut and the mono-

Switch operator positions	Switch	Switch
Switch	Switch	Switch
Switch	Switch	Switch
Toll room	Battery and rectifier room	
		Generator room

Figure 1.2 Generic MTSO configuration.

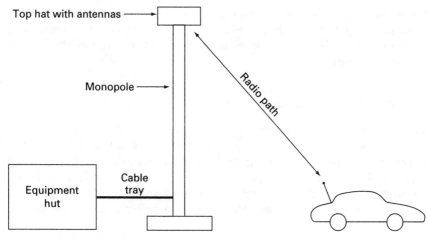

Figure 1.3 Monopole cell site.

pole supports the coaxial cables which connect the antennas to the radio-transmission equipment.

The radio-transmission equipment used for a cellular base station, located in the equipment room, is shown in Fig. 1.4. The equipment room layout is a typical arrangement in a cell site. The cell site radio equipment consists of a base site controller (BSC), radio bay, and the amplifier (TX) bay. The cell site radio equipment is connected to the antenna interface frame (AIF), which provides the receiver and transmit filtering. The AIF is then connected to the antennas on the monopole through use of the coaxial cables located next to the AIF bay.

The cell site is also connected to the MTSO through the Telco bay. The Telco bay provides either the T1 leased line or microwave radio link connection. The power for the cell site is secured through use of

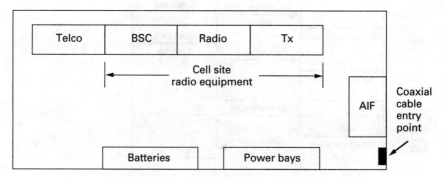

Figure 1.4 Generic cell site configuration. (Key: BSC = base site controller, radio = cellular radios, TX = amplifier, AIF = antenna interface, Telco = T1 microwave interconnect.)

power bays, rectifiers which convert ac electricity to dc. Batteries are used in the cell site in the event of a power disruption to ensure that the cell site continues to operate until power is restored or the batteries are exhausted.

1.5 Call Setup Scenarios

Several general call scenarios pertain to all cellular systems (Figs. 1.5 to 1.7). There currently are a few perturbations of the call scenarios discussed here driven largely by fraud-prevention techniques employed by individual operators. Numerous algorithms utilized throughout the call setup and processing scenarios are not included in the diagrams presented. However, the call scenarios presented

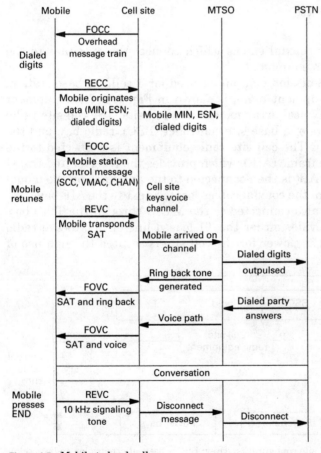

Figure 1.5 Mobile to land call.

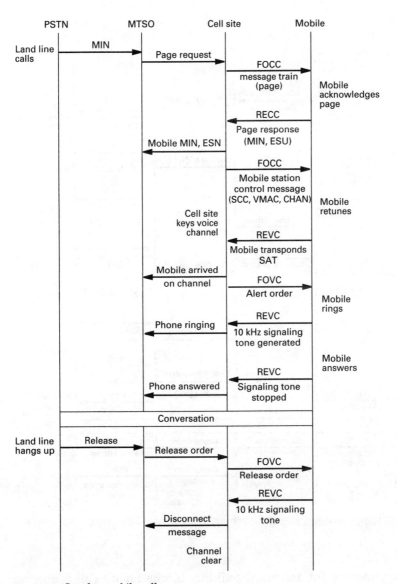

Figure 1.6 Land to mobile call.

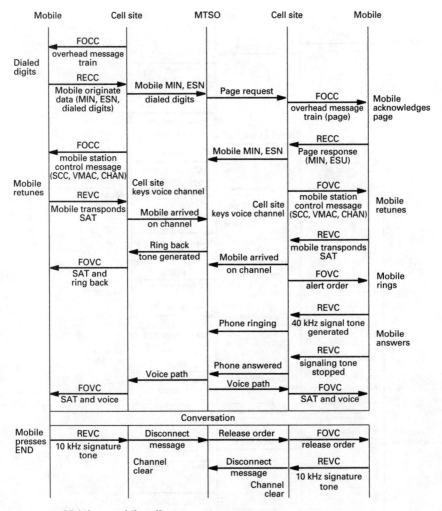

Figure 1.7 Mobile to mobile call.

here provide the fundamental building blocks for all call scenarios utilized in cellular systems.

1.6 Handoff

The handoff concept is one of the fundamental principles of this technology. Handoffs enable cellular systems to operate at lower power levels and provide high capacity. A multitude of algorithms are

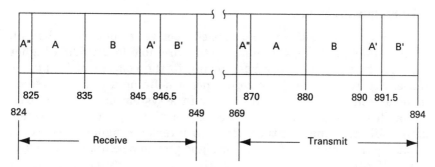

Figure 1.8 U.S. cellular spectrum chart. (All frequencies in MHz.)

invoked for the generation and processing of a handoff request and eventual handoff order. The individual algorithms are dependent upon the individual vendor for the network infrastructure and the software loads utilized.

Handing off from cell to cell is fundamentally the process of transferring the mobile unit that has a call in progress on a particular voice channel to another voice channel, all without interrupting the call. Handoffs can occur between adjacent cells or sectors of the same cell site. The need for a handoff is determined by the actual quality of the RF signal received from the mobile into the cell site. As the mobile unit traverses the cellular network, it is handed off from one cell site to another cell site, ensuring a quality call is maintained for the duration of the conversation.

1.7 Spectrum Allocation

Cellular systems have been allocated a designated frequency spectrum to operate within. Both A-band and B-band operators are allowed to utilize a total of 25 MHz of radio spectrum for their systems. The 25 MHz is divided into 12.5 MHz of transmit frequencies and 12.5 MHz of receive frequencies for each operator. The cellular spectrum is shown in Fig. 1.8. The location of the A-band and B-band cell site transmit and receive frequencies is indicated. Currently a total of 832 individual FCC channels are available in the United States. The radio channels utilized in cellular systems are spaced at 30-kHz intervals with the transmit frequency operating 45 MHz above the receive frequency. Both A-band and B-band operators have available a total of 416 radio channels, 21 setup and 395 voice channels.

1.8 Frequency Reuse

Cellular systems have a higher capacity per geographic area than an MTS or IMTS system. Cellular systems are able to provide a higher capacity to the subscriber base owing to the concept of frequency reuse. Frequency reuse, the core concept defining a cellular system, involves reusing the same frequency in a system many times over. The ability to reuse the same radio frequency many times in a system is a result of managing the carrier to interferer signal levels (C/I). Typically the minimum C/I level designed for in a cellular analog system is 17 dB.

Several frequency reuse patterns are currently in use throughout the cellular industry. Each of the different frequency reuse patterns has its advantages and disadvantages. The most common frequency reuse patterns employed in cellular systems are the $N = 4$ and $N = 7$ patterns. The number associated with the reuse pattern is an indication of the frequency repeat pattern used.

The frequency repeat pattern defines the maximum number of radios that can be assigned to an individual cell site. The $N = 7$ pattern can assign a maximum of 56 channels to a given cell site while the $N = 4$ enables 98 channels to be assigned. The $N = 4$ frequency pattern employs a cell site design that uses six sectors, and the $N = 7$ pattern uses a three-sector design. The difference between the three- and six-sector layouts is shown in Fig. 1.9.

1.9 Personal Communication Services (PCS)

Personal communication services (PCS) is the next generation of wireless communications. PCS has similarities and differences with

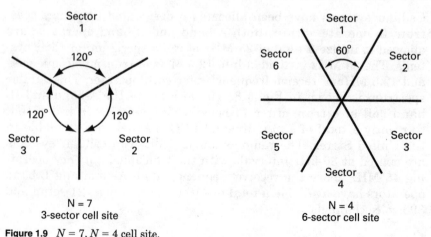

N = 7
3-sector cell site

N = 4
6-sector cell site

Figure 1.9 $N = 7, N = 4$ cell site.

the current cellular technologies. The similarities between PCS and cellular systems lie in the mobility of the user of the service. The differences between PCS and cellular fall into the applications and spectrum available for PCS operators to provide to the subscribers.

The PCS spectrum in the United States is being made available through an action process set up by the Federal Communications Commission (FCC). At the time of this writing both the A and B licenses for PCS have been awarded. The license breakdown is shown in Fig. 1.10.

The geographic boundaries for PCS licenses are different from those imposed on cellular operators in the United States. Specifically PCS licenses are defined as MTAs and BTAs. The MTA (Major Trading Area) has several BTAs (Basic Trading Areas) within its geographic region. A total of 93 MTAs and 487 BTAs are defined in the United States. Therefore, a total of 186 MTA licenses are awarded for the construction of a PCS network, each with a total of 30 MHz of spectrum to utilize. In addition a total of 1948 BTA licenses will be awarded in the United States. Of the BTA licenses the C band will have 30 MHz of spectrum while the D, E, and F blocks will have only 10 MHz available.

Currently there is no one standard for PCS operators to utilize for picking a technology platform for their networks. The choice of PCS standards is daunting, and each has its advantages and disadvantages. The current philosophy in the United States is to let the market decide which standard or standards are the best. This is significantly different from cellular systems, where every operator had a standard interface for the analog system from which to operate.

A few major standards have been picked by the licensees for the A and B PCS operators. The standards so far selected for PCS are DCS-1900, IS-95, IS-661, and IS-136. DCS-1900 utilizes a GSM format and is an upbanded DCS-1800 system. IS-95 is the CDMA standard that will be utilized by cellular operators, except upbanded to the PCS spectrum. IS-661 is a time-division duplex system offered by

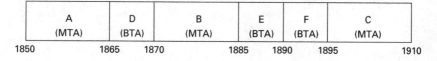

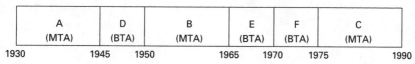

Figure 1.10   PCS spectrum allocation. (All frequencies in MHz.)

Omnipoint Communications. The IS-136 standard is an upbanded cellular TDMA system that is currently being used by cellular operators.

References

1. Jakes, *Microwave Mobile Communications,* IEEE Press, New York, 1974.
2. Lee, W. C. Y., *Mobile Cellular Telecommunications Systems,* McGraw-Hill, New York, 1989.
3. MacDonald, "The Cellular Concept," *Bell System Technical Journal,* vol. 58, no. 1, 1979.

Radio Engineering Topics

The rapid expansion of wireless technology in the marketplace has resulted in the deployment of a plethora of radio systems, and many new engineers are being thrust into the wireless arena. Much reference material is currently available for an RF engineer to learn from with respect to radio engineering. An excellent source is the textbooks used in engineering communication classes. However, the material presented in them focuses on the theory and not the practical aspects needed for operating a wireless system.

This chapter goes over radio principles incorporating the basic building blocks of a radio communication system. Its focus is to provide a brief discussion regarding some radio engineering principles. The radio principles covered in this chapter are not all-encompassing, since the field of radio engineering is very vast. However, the intention is to cover the most relevant issues that are needed as a basic foundation for a wireless engineer to know.

2.1 The Radio System

The fundamental building blocks of a communication system are shown in Fig. 2.1. The simplified drawing represents the major components in any communication system, consisting of an antenna, filters, receivers, transmitter, modulation, demodulation, and propagation. Entire books can be and have been written on each of the components listed. Knowing their design characteristics is essential in building a communication system that will provide the proper transport functions for the information content.

At the end of the chapter is a reference section that I have found to be very useful. It is only a brief listing of the many sources of infor-

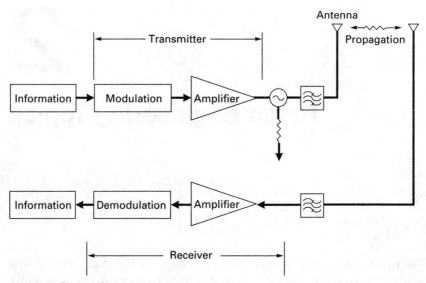

Figure 2.1 Basic radio system.

mation available on radio engineering, but it serves as a good starting point for the reader.

2.2 Propagation

The ability of RF energy to propagate through the air is a fundamental concept and principle which enables wireless companies to deliver a service without physically connecting devices to the network. RF energy is comprised of an electromagnetic energy, best described by Fig. 2.2. The frequency of oscillation of the electromagnetic wave is directly related to the physical wavelength, as shown in Eq. (2.1).

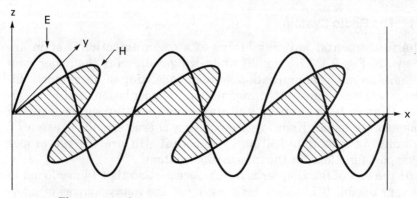

Figure 2.2 Electromagnetic wave.

$$C = f\lambda \qquad (2.1)$$

where f = frequency, Hz
$c = 3\times10^8$ m/s
λ = wavelength, m

Some quick manipulation of Eq. (2.1) results in establishing the wavelength for a cellular base station transmitting site. The frequency selected for this example is 880 MHz, since it is right in the middle of both the A- and B-band users.

$$\lambda = \frac{c}{f} = \frac{3\times10^8 \text{ m/s}}{880\times10^6 \text{ Hz}} \qquad (2.2)$$

where $\lambda = 0.34$ m ($\cong 13$ in).

2.2.1 Propagation model

The radio wavelength for cellular systems is rather small in size and as a result has some unique propagation characteristics which have been modeled by numerous technical people over the years. Some of the more popular propagation models used are Hata, Carey, Longley-Rice, Bullington, and Cost231. Each of these models has advantages and disadvantages. Specifically some baseline assumptions used with any propagation model need to be understood before they are used. Most cellular operators use a version of the Hata model for conducting propagation characterization. The Carey model, however, is used for submitting information to the FCC with regard to cell site filing information.

The model used for predicting coverage needs to have factored in to it many variables which directly impact the actual RF coverage prediction of the site. The positive attributes affecting coverage are the receiver sensitivity, transmit power, antenna gain, and antenna height above average terrain. The negative factors affecting coverage involve line loss, terrain loss, tree loss, building loss, electrical noise, natural noise, antenna pattern distortion, and antenna inefficiency, to mention a few.

With the proliferation of cell sites the need to theoretically predict the actual path loss experienced in the communication link is becoming more and more critical. To date no overall theoretical model has been established that explains all the variations encountered in the real world. However, as the cellular communication systems continue to increase a growing reliance is placed on the propagation prediction tools. This reliance is intertwined in the daily operation of the cellular communication system. The propagation model employed by the cellular operator is directly involved with the capital build program of the company for determining the budgetary requirements for the next few fiscal years.

Therefore, it is essential that the model utilized for the propagation prediction tool be understood. It should be understood in terms of what it can and cannot predict. Over the years numerous articles have been written with respect to propagation modeling in the cellular communications environment. With the introduction of personal communication services (PCS) there has been increased focus on refining the propagation models to assist in planning the networks.

Presently most available cellular propagation tools utilize a variation of the Hata model.[18] The potential PCS and some cellular operators are utilizing the Cost231 model.[35] However, there are several variants to the Cost231 model as there are with the Hata model. Which model should be used, its baseline assumptions, and its performance relative to the others are briefly covered here. As mentioned before, several propagation models are currently utilized throughout the industry. Each has pros and cons. Through understanding their advantages and disadvantages, a better engineering design can take place in a network. The specific models discussed here are free space, Hata, and the Cost231 model.

Free space.   Free space path loss is usually the reference point for all the path loss models employed. Each propagation model points out that it more accurately predicts the attenuation experienced by the signal over that of free space. The equation used for determining free space path loss is based on $1/R^2$ or 20 dB per decade path loss. It is shown in Eq. (2.3):[35]

$$L_f = 32.4 + 20 \log_{10} R + 20 \log_{10} f_c \qquad (2.3)$$

where R = distance from cell site, km
f_c = transmit frequency, MHz
L_f = free space path loss, dB

The free space path loss equation has a constant value that is used for the air interface loss, a distance, and frequency adjustments. Using some basic values the different path loss values can be determined for comparison with later models discussed. The baseline assumptions in Table 2.1 are distances in kilometers and a frequency of 880 MHz.

Looking at the table, it would be very nice if the frequency band utilized by cellular operators behaved with this path loss. However, cellular frequency propagation does not behave like free space loss and therefore requires another equation.

TABLE 2.1

Distance, km	Path loss, dB
1.0	91.29
2.0	97.31
3.0	100.83
4.0	103.33
5.0	105.27

The Hata model. The most prolific path loss model employed in cellular systems presently is the empirical model developed by Hata or some variant of it. The Hata model is an empirical model derived from the technical report made by Okumura so the results could be used in a computational model. The Okumura report[38] is a series of charts that are instrumental in radio communication modeling. The Hata model is shown in Eq. (2.4):[18]

$$L_H = 69.55 + 26.16 \log_{10} f_c - 13.82 \log_{10} h_b$$
$$- a(h_m) + (44.9 - 6.55 \log_{10} h_b) \log_{10} R \qquad (2.4)$$

where L_H = path loss for Hata model
   h_b = base station antenna height
   h_m = mobile antenna height

Utilizing the same values used for the free space calculation, a similar chart is derived. It should be noted that some additional conditions are applied when using the Hata model. The values are dependent upon the range over which the equation is valid. If the equation is used with parameters outside the values for which it is defined, the results will be suspected of error. The range for which the Hata model is valid is listed below.

$$f_c = 150 \text{ to } 1500 \text{ MHz}$$

$$h_b = 30 \text{ to } 200 \text{ m}$$

$$h_m = 1 \text{ to } 10 \text{ m}$$

$$R = 1 \text{ to } 20 \text{ km}$$

Therefore, the Hata model should not be employed when trying to predict path loss less than 1 km from the cell site or if the site is less than 30 m in height. This is an interesting point to note, since cellular sites are being placed at times less than 1 km apart and often below the 30-m height.

In the Hata model the value $a(h_m)$ is used to correct for the mobile antenna height. The interesting point is that if you assume a height of 1.5 m for the mobile, that value nulls out of the equation. Assume the conditions shown for the Hata model.

$$F_c = 880 \text{ MHz}$$

$$h_b = 30 \text{ m}$$

$$h_m = 1.5 \text{ m}$$

A critical point to mention here is that the Hata model employs three correction factors based on the environmental conditions that path loss prediction is evaluated over. The three environmental conditions are urban, suburban, and open. For Table 2.2 I have assumed an open, rural, environment to best compare the two equations.

The environmental correction values are easily calculated but vary for different values of mobile height. For the values listed below a mobile height of 1.5 m was assumed.

Environmental correction	Condition factor, dB
Urban	0
Suburban	−9.88
Open	−28.41

Cost231 Walfisch/Ikegami. The Cost231 Walfisch/Ikegami propagation model[35] is used for estimating the path loss in an urban environment for cellular communication. The Cost231 model is a combination of empirical and deterministic modeling for estimating the path loss in an urban environment over the frequency range of 800 to 2000 MHz.

TABLE 2.2   Conditions Assumed for the Hata Model

Distance, km	Path loss, dB		
	Free space	Hata (rural)	Hata (urban)
1.0	91.29	97.75	126.16
2.0	97.31	108.36	136.77
3.0	100.83	114.56	142.97
4.0	103.33	118.96	147.37
5.0	105.27	122.37	150.79

The Cost231 model is used primarily in Europe for GSM modeling and in some propagation models used for cellular systems in the United States.

The Cost231 model is composed of three basic components:

1. Free space loss
2. Roof to street diffraction loss and scatter loss
3. Multiscreen loss

$$L_C = \text{where } L_{\text{rts}} + L_{\text{ms}} \le 0 \quad \frac{L_f}{L_f} + L_{\text{rts}} + L_{\text{ms}} \tag{2.5}$$

where L_f = free space loss
$\quad L_{\text{rts}}$ = rooftop to street diffraction and scatter loss
$\quad L_{\text{ms}}$ = multiscreen loss

$$L_f = 32.4 + 20 \log_{10} R + 20 \log_{10} f_c \tag{2.6}$$

$$L_{\text{rts}} = -16.9 - 10 \log_{10} w + 10 \log f_c + 20 \log \Delta h_m + L_0 \tag{2.7}$$

where w = street width in meters and $\Delta h_m = h_r - h_m$

$$L_0 = \begin{cases} -10 + 0.354 & 0 \le \theta \le 35 \\ 2.5 + 0.075\,(\theta - 35) & 35 \le \theta \le 55 \\ 4.0 - 0.114\,(\theta - 55) & 55 \le \theta \le 90 \end{cases} \tag{2.8}$$

where θ = incident angle relative to the street
$\quad L_0$ = correction factor for street orientation

$$L_{\text{ms}} = L_{\text{bsh}} + k_a + k_d \log_{10} R + k_p \log_{10} f_c - 9 \log b \tag{2.9}$$

where b = distance between buildings along the radio path, m
$\quad L_{\text{bsh}}$ = path loss correction for base station antenna height relative to the roof, m
$\quad K_a$ = path loss correction factor for increased path loss for lower antenna heights
$\quad K_d$ = path loss correction factor for multiscreen diffraction loss versus distance
$\quad K_f$ = path loss correction factor for multiscreen diffraction loss versus frequency

$$L_{bsh} = \begin{cases} -18 \log_{10}(1 + \Delta h_b) & h_b > h_r \\ 0 & h_b < h_r \end{cases} \qquad (2.10)$$

$$K_a = \begin{cases} 54 & h_b > h_r \\ 54 - 0.8\, h_b & d \geq 500 \text{ m}, h_b \leq h_r \\ 54 - 1.6(\Delta h_b)R & d < 500 \text{ m}, h_b \leq h_r \end{cases} \qquad (2.11)$$

Note: Both L_{bsh} and K_a increase the path loss with lower base station antenna heights.

$$K_d = \begin{cases} 18 & h_b > h_r \\ 18 - 15 \dfrac{\Delta h_b}{\Delta h_r} & h_b \leq h_r \end{cases} \qquad (2.12)$$

$$K_f = \begin{cases} 4 + 0.7\left(\dfrac{f_c}{925} - 1\right) \\ 4 + 1.5\left(\dfrac{f_c}{925} - 1\right) \end{cases} \qquad (2.13)$$

The top value for K_f is for a midsized city and suburban area with moderate tree density. The second line is for a metropolitan center.

Equations (2.5) to (2.13) compose the Cost231 model, and the following items bound the equations' useful range. It is important, as always, to know what the valid ranges are for the model. The following valid ranges for the cost 231 model

$$f_c = 800 \text{ to } 2000 \text{ MHz}$$

$$h_b = 4 \text{ to } 50 \text{ m}$$

$$h_m = 1 \text{ to } 3 \text{ m}$$

$$R = 0.02 \text{ to } 5 \text{ km}$$

show that when the range of the site is less than 1 km the Cost231 model would make a better choice than the Hata model.

Some additional default values apply to the Cost231 model when specific values are not known. The default values recommended are listed below. They can and will significantly alter the path loss values arrived at.

$$b = \text{distance between buildings } (= 20 \text{ to } 50 \text{ m})$$

$$w = \text{width of street m} \left(= \frac{b}{2} \right)$$

$$h_r = \text{height to roof, m [3 (number of floors) + roof]}$$

$$\text{Roof} = \begin{cases} 3 \text{ m} & \text{for pitched roof} \\ 0 \text{ m} & \text{for flat roof} \end{cases}$$

$$\theta = \text{incident angle } (= 90°)$$

Assuming the following values for the Cost231 and the same values used for the previous equations for free space and the Hata path loss models, another comparison is presented in Table 2.3.

$$f_c = 880 \text{ MHz}$$

$$h_m = 1.5 \text{ m}$$

$$h_b = 30 \text{ m}$$

$$\text{Roof} = 0 \text{ m}$$

$$h_r = 30 \text{ m}$$

$$\theta = 90°$$

$$b = 30 \text{ m}$$

$$w = 15 \text{ m}$$

Obviously the assumptions made for entering values into the equations play a major role in defining the outcome of the path loss value.

Figure 2.3 is a plot of the propagation models referenced in this

TABLE 2.3

Distance, km	Path loss, dB		
	Free space	Hata (urban)	Cost231
1.0	91.29	126.16	139.45
2.0	97.31	136.77	150.89
3.0	100.83	142.97	157.58
4.0	103.33	147.37	162.33
5.0	105.27	150.79	166.01

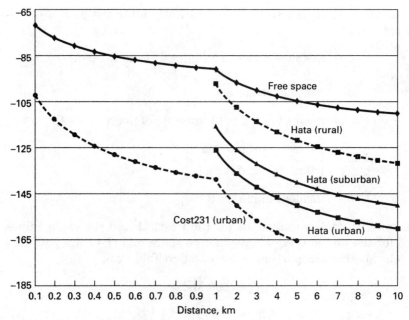

Figure 2.3 Propagation model comparison.

chapter for several environments. The plot shows the similarity and differences between models utilizing the same, or near same, parameters. The assumptions made for the models employed in the graph (Fig. 2.3) utilize the following baseline conditions:

$$f_c = 880 \text{ MHz}$$

$$h_m = 1.5 \text{ m}$$

$$h_b = 30 \text{ m}$$

$$\text{Roof} = 0 \text{ m}$$

$$h_r = 30 \text{ m}$$

$$b = 30 \text{ m}$$

$$w = 15 \text{ m}$$

Figure 2.4 is a graphical illustration of the various Cost231 parameters.

The propagation model or models employed by your organization must be chosen with extreme care and undergo a continuous vigil to ensure they are truly being a benefit to the company as a whole. The propagation model employed by the engineering department not only determines the capital build program but also plays a direct factor in the performance of the network.

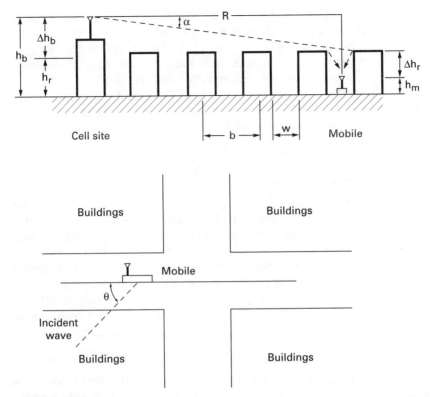

Figure 2.4 Diagram.

The capital build program is directly affected by the propagation model chosen. The model is normally used as part of the 1-, 2-, and 5-year growth study performed by a cellular operator. It is used to determine how many sites are needed to provide a particular coverage requirement needed for the network. In addition to the coverage requirement are traffic loading requirements which rely on the propagation model chosen to determine the traffic distribution, offloading, from existing site to new sites as part of the capacity relief program. The propagation model helps determine where the sites should be placed to achieve an optimal position in the network. If the propagation model used is not effective in helping place sites correctly, the probability of incorrectly justifying and deploying a site into the network is high.

The performance of the network is also affected by the propagation model chosen, since it is used for interference prediction plots. If the propagation model is inaccurate by say 6 dB, you could be designing for a 23- or 11-dB C/I. Based on your traffic loading conditions, designing for a very high C/I level could have a negative financial impact. On the other hand, designing for a low C/I would have the obvious impact of degrading the quality of service to the very people who pay your salary. The propagation model is also used in a multitude of other system performance aspects that include handoff tailoring, power-level adjustments, and antenna placements, to mention a few.

Reiterating the point that no model can account for all the perturbations experienced in the real world, it is essential that you utilize one or several propagation models for determining the path loss of your network.

2.2.2 Environmental attenuation

Focusing back on one of the more popular propagation models, the Hata model, it uses an alpha of about 3.5. There have been many debates regarding the propagation tools of dB per decade slope used. However, it is the other variables associated with the propagation characteristics that have the largest impact on how the model truly predicts the propagation characteristics of the potential cell site.

The difference between alphas of 3.5 and 3.8 is only 3 dB per decade. The 3 dB can easily be aborted in environmental attenuation issues. The environmental issues associated with propagation characteristics require the highest amount of focus. Table 2.4 lists some generalized attenuation characteristics used in propagation models. The difference between the urban and suburban environments is more than 3 dB. Obviously each area has its own unique propagation characteristic, and generalizing the characteristics of an area is a best-guess fit. The table is meant to be used as a general guide for deter-

TABLE 2.4 Attenuation Table

Environmental loss, dB	
Foliage loss:	
Sparse	6
Light	10
Medium	15
Dense	20
Very dense	25
Building loss:	
Water/open	0
Rural	5
Suburban	8
Urban	22
Dense urban	27
Vehicle	10–14

mining environmental effects on path loss. However, looking at Table 2.4 it is obvious that 3 dB per decade is not the leading indicator for modeling propagation.

2.2.3 Diffraction

Diffraction of the RF signal also has a very important role in predicting and attenuating the signal. How to calculate the actual attenuation from diffraction is shown in an excellent paper by Bullington.[7] Several types of diffraction methods are modeled in RF; they are smooth and knife-edge. Each diffraction method yields a different value. If you choose wisely for the terrain issues at hand, the calculation method presented will accurately predict the attenuation experienced. The differences in attenuation involving antenna heights and a knife-edge diffraction point are shown in Table 2.5.

One good use for knowing how to calculate diffraction is for trying to determine the loss of signal expected in a major valley. Another is calculating the signal loss expected by using a mountain for containing the signal through placement of the antenna. Still another is

TABLE 2.5 Diffraction Loss Table

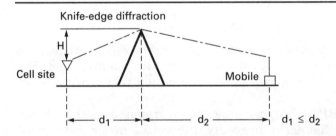

| H | d_1 | | |
	Feet	Meters	dB loss
50 ft (15 m)	50	15	23
	100	30	21
	200	60	17
	500	150	13
100 ft (30 m)	50	15	38
	100	30	37
	200	60	35
	500	150	31
200 ft (60 m)	50	15	47
	100	30	44
	200	60	40
	500	150	36

reducing the tilt angle of a cell site and estimating the positive and negative impact of improving the signal over a ridge or in a valley.

2.3 Effective Radiated Power (ERP)

The actual effective radiated power (ERP) for the cell site or transmitter used for the communication site will determine its transmit radius. The ERP setting should be balanced with the receive path to ensure that there is no disparity between talk-out (transmit) and talk-back (receive) paths for a cell site. The ERP for the site is set with reference to a dipole antenna. The method for calculating the ERP for a site is as follows, using the data provided in the chart below.

Transmitter output	44 dBm (25 W)
Combining losses	−1 dB
Feedline loss	−3 dB
Antenna gain	10 dBd
	50 dBm (100 W)

If you are using dBi (isotropic gain), or want to convert dBi to dBd, all that is involved is a simple conversion:

$$\text{dBd} = 2.14 \text{ dB} + \text{dBi}$$

The ERP for the site has a dramatic impact on the cell site radius. An increase in ERP could reduce the build program requirements. The change to the build program is best explained through use of an example.

Altering the ERP by 3 dB can alter the geographic area the cell site serves, assuming flat earth. In reality if there is a major obstruction, increasing the ERP may have little or no real effect, depending upon the actual terrain conditions. However, referring back to the propagation model which has a slope of 35 dB per decade, this value is used to determine the effective cell radius (ECR). Example 2.1 evaluates a 3-dB increase and decrease in ERP and its effect on the ECR.

Example 2.1

$$\text{Area} = \pi R^2$$

3-dB increase in ERP:

$$+ 3 \text{ dB} = 35 \log (R_{\text{new}})$$

$$\therefore R_{\text{new}} = 1.218 \qquad R_{\text{old}} = 1.0$$

$$\frac{\text{Area (new)}}{\text{Area (old)}} = \frac{\pi R^2_{\text{new}}}{\pi R^2_{\text{old}}} = \frac{R^2_{\text{new}}}{R^2_{\text{old}}} = R^2_{\text{new}} = 1.484$$

$$\text{Area (new} + 3 \text{ dB)} = 1.484 \text{ Area}_{old}$$

3-dB decrease in ERP:

$$-3 \text{ dB} = 35 \log (R_{new})$$

$$\therefore R_{new} = 0.820 \qquad R_{old} = 1.0$$

$$\frac{\text{Area (new)}}{\text{Area (old)}} = \frac{\pi R_{new}^2}{\pi R_{old}^2} = \frac{R_{new}^2}{R_{old}^2} = R_{new}^2 = 0.674$$

$$\text{Area (new} -3 \text{ dB)} = 0.674 \text{ Area}_{old}$$

The actual effect the ERP has on the build program is best shown in Example 2.2.

Example 2.2   Total geographic area to cover = 50 km².

$$\text{Area (old)} = \pi R_{old}^2 = 3.14 \text{ km}^2 \qquad R_{old} = 1 \text{ km}$$
$$\text{Area (+ 3 dB)} = \pi R_{new}^2 = 4.66 \text{ km}^2 \qquad R_{new\ +\ 3\ dB} = 1.484 \text{ km}$$
$$\text{Area (}-3 \text{ dB)} = \pi R_{new}^2 = 2.16 \text{ km}^2 \qquad R_{new\ -3\ dB} = 0.674 \text{ km}$$

Approximate number of cells needed to cover 50 km²:

$$\text{Number of cell sites} = \frac{\text{geographical area}}{\text{cell site area}}$$

$$\text{Number of cell sites (old)} = \frac{50 \text{ km}^2}{3.14 \text{ km}^2} \approx 16$$

$$\text{Number of cell sites (new } + 3 \text{ dB)} = \frac{50 \text{ km}^2}{4.66 \text{ km}^2} \approx 11$$

$$\text{Number of cell sites (new } -3 \text{ dB)} = \frac{50 \text{ km}^2}{2.16 \text{ km}^2} \approx 23$$

This simple example shows that a mere 3-dB increase in ERP across all the cells in the network results in a decrease of 31 percent in the total number of cell sites required. Conversely a 3-dB reduction in ERP increases the cell site build program by 43 percent. The ERP should not be increased or decreased by itself without a careful analysis of the link budget for the communication system.

2.4   Link Budget

The maximum path loss, or limiting path, for any communication system used determines the effective range of the system. Table 2.6 involves a simple calculation of a link budget for the determination of

TABLE 2.6   Link Budget

	Downlink	Uplink
Transmit	50 dBm	36 dBm
Antenna gain	3 dBd	12 dBd
Cable loss	2 dB	3 dB
Receiver sensitivity	−116 dBm	−116 dBm
C/N ratio	17 dB	17 dB
Maximum path loss	150 dB	144 dB

which path is the limiting case to design from. The thermal noise, bandwidth, and noise figures are factored into the receiver sensitivity value presented. The uplink path, defined as mobile to base, is the limiting path case. From Table 2.6 the talk-back path is 6 dB less than the talk-out path. The limiting path loss is then used to determine the range for the site using the propagation model for the network.

2.5   Antennas

The antennas used by the cell site and mobile for establishing and maintaining the communication link are a crucial element of the system. The antenna system is the interface between the radio system and the external environment. Many types of antennas are available, all of which perform specific functions depending on the application at hand. The type of antenna used by a system operator can be a collinear, log-periodic, folded-dipole, or yagi, to mention a few.

The two primary classifications of antennas for a system are omni and directional. Omni antennas are used when the desire is to obtain a 360° radiation pattern. Directional antennas are used when a more refined pattern is desired. The directional pattern is usually needed to facilitate system growth through frequency reuse or to shape the system's contour. The antenna you use for a network should match the system design objectives.

2.5.1   Types

Many types of antennas are available for use on the commercial market without the need to invent more. Two common types of antennas used in cellular communication systems are collinear and log-periodic antennas. Collinear antennas can be either omni or directional. They operate with a series of dipole elements that operate in phase and is referred to as a broadside radiator. The maximum radiation for the collinear antenna takes place along the dipole arrays axis, and the array consists of a number of parallel elements in one plane.

The other general type of antenna used is a log-periodic dipole array (LPDA), a directional antenna whose gains, standing-wave ratio (SWR), and other key figures of merit remain constant over the operating band. The LPDA is used where a large bandwidth is needed and the typical gain is 10 dBi.

The LPDA has a structural geometry such that its impedance and radiation characteristics repeat periodically as the log of the frequency. The actual antenna consists of several dipole elements which have different lengths and different relative spacing. Figure 2.5 shows an LPDA antenna. At the frequency of interest the short elements take on a capacitance property and receive little power by acting as a parasitic director. The longer elements, however, take on an inductive characteristic and are more dominant, acting as parasitic reflectors.

2.5.2 Figure of merit

Many parameters and figures of merit characterize the performance of an antenna system. The following is a partial list of the figures of merit for an antenna; it should be quantified by the manufacturer of the antennas you are using.

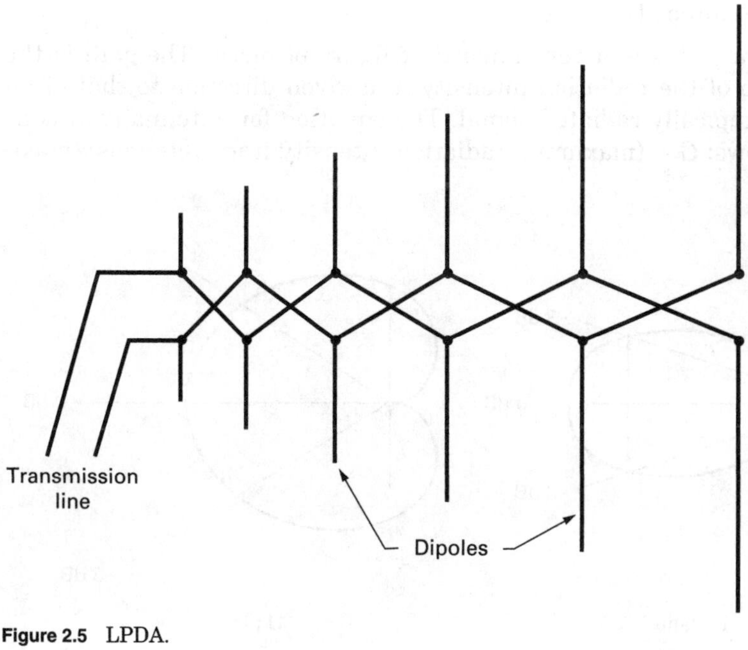

Transmission
line

Dipoles

Figure 2.5 LPDA.

1. *Antenna pattern.*   This is the graphical representation of the antenna pattern.

2. *Main lobe.*   This is the radiation lobe containing the direction of maximum radiated power.

3. *Side lobe.*   This is the radiation's lobe in any direction other than the main lobe.

4. *Input impedance.*   This is the impedance presented by the antenna at its terminals and is usually complex.

5. *Radiation efficiency.*   This is the ratio of total power radiated by an antenna to the net power accepted by an antenna from the transmitter. The equation is as follows: e = (power radiated)/(power radiated + power lost). The antenna would be 100 percent efficient if the power lost in the antenna were zero.

6. *Beamwidth.*   This is the angular separation between two directions in which radiation interest is identical. The half-power point for the beamwidth is usually the angular separation where there is a 3-dB reduction off the main lobe (Fig. 2.6).

7. *Directivity.*   This is the ratio of radiation intensity in a given direction to that of the radiation intensity averaged over all the other directions. The equation for antenna directivity is as follows: $G(D)$ = (maximum power radiation intensity)/(average radiation intensity).

8. *Gain.*   This is a very important figure of merit. The gain is the ratio of the radiation intensity in a given direction to that of an isotropically radiated signal. The equation for antenna gain is as follows: G = (maximum radiation intensity from antennas)/(maxi-

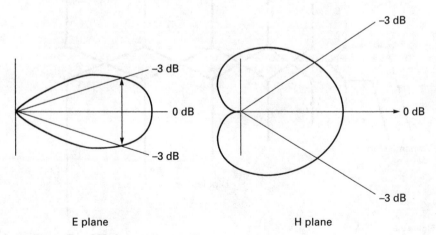

E plane   H plane

Figure 2.6   E and H plane antenna patterns.

mum radiation from an isotropic antenna). The gain of the antenna can also be described as $G = e \times G(D)$ and if the antenna were without loss, $e = 1$, then $G = G(D)$.

9. *Antenna polarization.*   The antenna polarization is defined as the polarization fields radiated by the antenna. The antenna's polarization is defined by the E field vector. Cellular systems use vertical polarization.

10. *Bandwidth.*   The bandwidth defines the operating range of the frequencies for the antenna. The SWR is usually how this is represented besides the frequencies range over which it is constant.

11. *Power dissipation.*   The total power the antenna can accept at its input terminals is its power dissipation.

12. *Intermodulation.*   The amount of intermodulation the antenna will introduce to the network in the presence of strong signals.

13. *Construction.*   The construction attributes associated with its physical dimensions, mounting requirements, materials used, wind loading, connectors, and color constitute this figure of merit.

14. *Cost.*   How much the antenna costs is a critical figure of merit.

2.6   Filters

Filters are an often forgotten aspect in a cellular communication system. However, the filters used in the network play a vital role in protecting the receiver from unwanted signals within a given cell site. Several types of filters are available and are used. These include bandpass, low-pass, high-pass, and notch-type filters (Fig. 2.7).

The low-pass filter is meant to pass only those frequencies which are below its cutoff frequency. The high-pass filter rejects all frequencies below its cutoff frequency and passes the rest. The bandpass filter passes frequencies within a specified band of frequencies and attenuates all the frequencies outside of the passband. The notch filter, or band reject filter, has the ability to highly attenuate specific frequencies within a specified band and pass the rest.

Cellular systems employ several types of filters in a normal communication path. The A-band carrier usually employs two bandpass filters in series as the front-end filter for the receive path. The B-band carrier usually employs a bandpass filter with a notch to obtain its frequency selectivity. The actual performance of the filters should be observed because the filter is meant to protect the receiver from unwanted signals. However, if the filter protects the receiver too much, excess attenuation occurs over frequencies desired to be passed

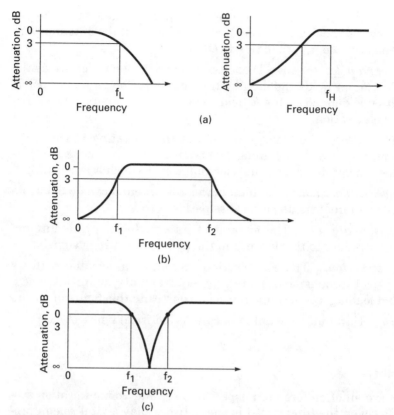

Figure 2.7 Filter types. (a) Low-pass filter and high-pass filter; (b) bandpass filter; (c) notch (band-reject) filter.

through. Figure 2.8 shows a typical bandpass filter response desired for the A-band carrier, and Fig. 2.9 is for the B-band carrier.

The filter used for the system's front end should meet several performance criteria. The first performance criterion is the passband ripple, the variation in attenuation (insertion loss) that the filter will inflict on a signal across the band. The passband ripple is important because if it is too great, a variation in performance will take place for a system on a channel-by-channel basis because of the filter itself. The passband ripple of a filter should be less than 1 dB.

The insertion loss created by a filter is also important since this is a measure of how much attenuation the filter imposes on frequencies to be passed through the filter itself. The insertion loss the filter imposes on the frequencies to be passed should be less than 1 dB. However, the attenuation at the band edge of the filter is critical.

Ideally at the band edge the filter should pass the last frequency desired with little or no attenuation and provide infinite attenuation

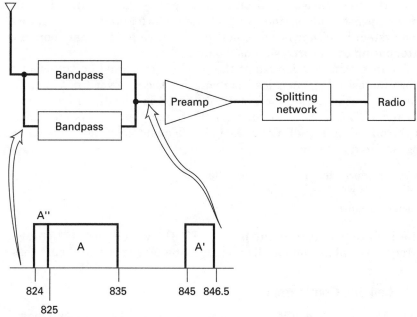

Figure 2.8 A-band receive.

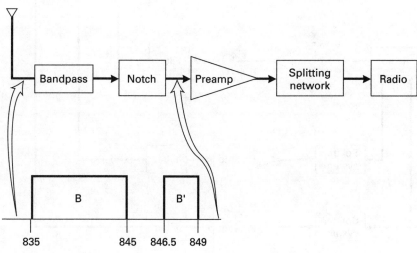

Figure 2.9 B-band receive.

to all the frequencies not desired to be passed. In practice the band edge response could seriously affect the performance of a communication system by either not attenuating the unwanted signal enough or attenuating the desired signals too much.

The insertion loss desired at the band edges should be sufficient to reject undesired frequencies from entering the receiver front end. The amount of attenuation, or isolation, needed is calculated by examining the 1-dB compression point for the preamp and the power output of the offending signal. Below is a simple example of how to arrive at the required isolation.

Offending transmitter ERP	50 dBm
1 dB compression point	-25 dB
Isolation required	75 dB

The isolation requirement here is 75 dB, which can be achieved through a combination of filtering, antenna placement, and path loss.

2.7   Cell Site Configurations

Numerous cell site configurations can take place. The scope and magnitude of the configuration variations are largely due to infrastructure vendor design variations and operator requirements. However, all the cellular sites have some common elements regardless of the vendor.

Figure 2.10 is a typical cell site radio path. It employs a total of

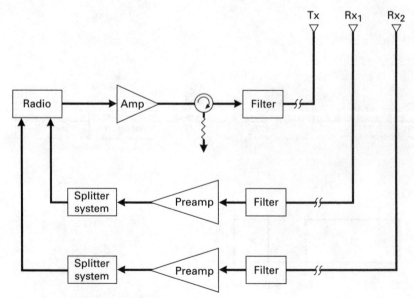

Figure 2.10   Cell site radio path.

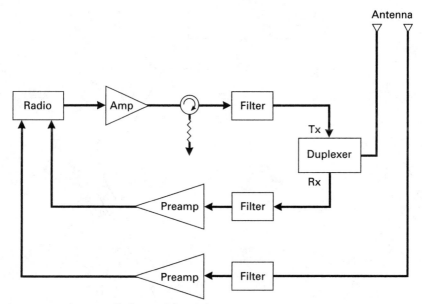

Figure 2.11 Antenna duplex configuration.

three antennas for the actual radio system, one to transmit and two to receive. The two receiving antennas are utilized for diversity and fade margin protection. The cell site radio configuration shown can be applied to an omni, three-sector, or six-sector cell site.

Sometimes it becomes necessary to reduce the number of antennas on an installation for a multitude of reasons ranging from cost to least constraints. Figure 2.11 shows an antenna duplex configuration. Figure 2.12 is a typical antenna configuration layout for a three-sec-

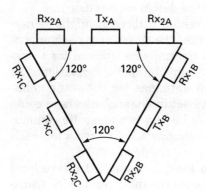

Figure 2.12 Three-sector antenna array.

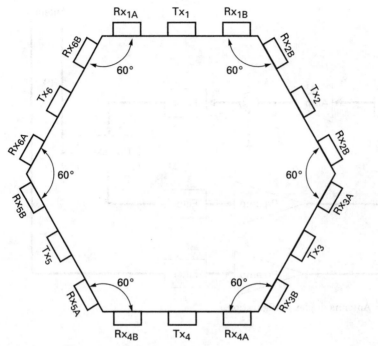

Figure 2.13 Six-sector antenna array.

tor cell site. The example shown employs a total of three antennas per sector, one transmit and two receive. The actual number of antennas utilized for the cell site configuration could be more, usually one additional transmit antenna, depending on the design requirements.

Figure 2.13 is a typical antenna configuration layout for a six-sector cell site. The example shown employs a total of three antennas per sector, one transmit and two receive. Again the actual number of antennas utilized for the cell site configuration could be more, usually one additional transmit antenna, depending on the design requirements.

Depending upon the infrastructure vendor utilized it might be necessary to employ omni antennas for either the three- or six-sector configuration. The omni antennas would normally be utilized for transmitting and receiving the control or setup channel information. For infrastructure vendors requiring omni antennas for the setup channels a disparity between the voice and setup channel coverage could take place. To alleviate the potential of having two separate communication paths for the voice and setup channels, a form of simulcasting was developed to eliminate this type of problem.

The simulcast designs presented in Figs. 2.14 and 2.15 have been used to distribute the setup channels onto the individual voice trans-

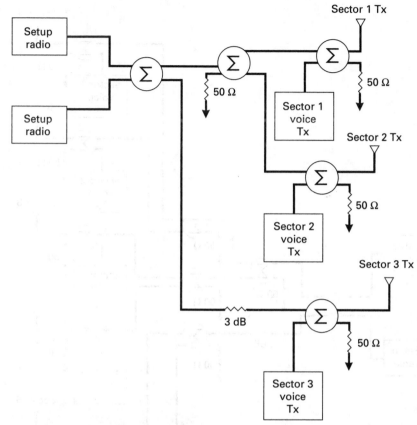

Figure 2.14 Three-sector setup on sector (SOS).

mit antennas. The configuration shown here references only the transmit path of the design. The receive path portion of the simulcast design is not shown, however. There are other methods of simulcasting the setup channel on the voice antennas.

2.8 Modulation and Demodulation

To convey voice and data information utilizing the electromagnetic wave it is necessary to modulate the carrier wave at the transmitting source and then demodulate it at the receiver. The generalized radio system is shown in Fig. 2.16. The choice of modulation and demodulation utilized for the radio communication system is directly dependent upon the information content desired to be sent, the available spectrum to convey the information, and the cost. The fundamental goal of modulating any signal is to obtain the maximum spectrum efficiency, or rather information density, per hertz.

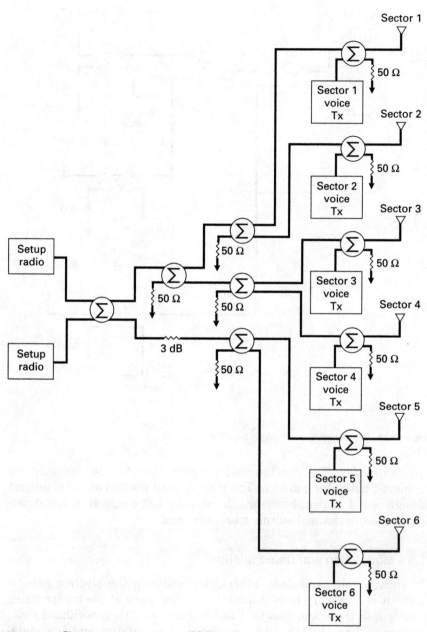

Figure 2.15 Six-sector setup on sector (SOS).

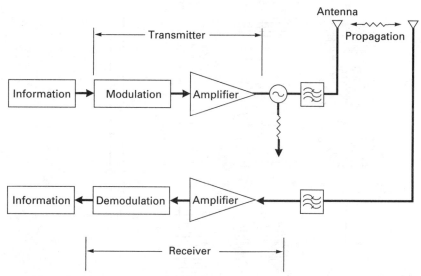

Figure 2.16 Basic radio system.

Many types of modulation and demodulation formats are utilized for the conveyance of information. However, all the communication formats rely on one, two, or all three of the fundamental modulation types. The fundamental modulation techniques are amplitude modulation (AM), frequency modulation (FM), and phase modulation (PM). Figure 2.17 highlights the differences between the modulation techniques in terms of their impact on the electromagnetic wave itself.

Amplitude modulation (AM) has many unique qualities. However, this form of communication is not utilized in cellular communication primarily because it is more susceptible to noise.

Frequency modulation (FM) is utilized for AMPS and TACS analog communication. FM is utilized for analog cellular communication since it is more robust to interference. Cellular systems utilize a channel bandwidth of 30 kHz for FM modulation.

Phase modulation (PM) is used for TDMA and CDMA communication systems. There are many variations to phase modulation. Specifically many digital modulation techniques rely on modifying the RF carrier phase and amplitude.

Quadrature phase shift keying (QPSK) is one form of digital modulation which has a total of four unique phase states to represent data. The four phase states are arrived at through different I and Q values, utilizing four phase states, i.e., quadrature, which allows each phase state to represent two data bits. The two data bits are mapped on the IQ chart in Fig. 2.18.

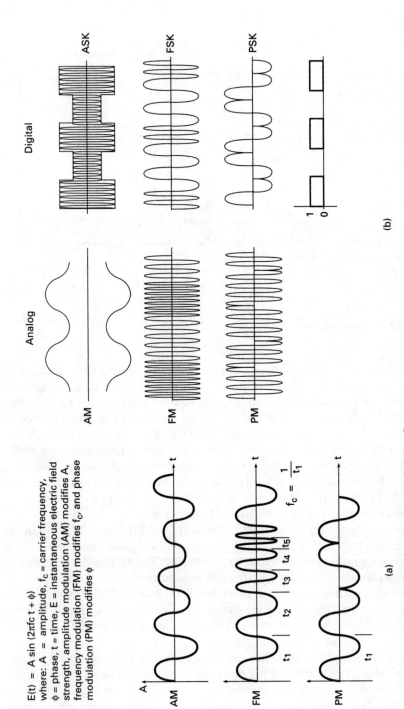

$E(t) = A \sin (2\pi f_c t + \phi)$

where: A = amplitude, f_c = carrier frequency,
ϕ = phase, t = time, E = instantaneous electric field
strength, amplitude modulation (AM) modifies A,
frequency modulation (FM) modifies f_c, and phase
modulation (PM) modifies ϕ

Figure 2.17 (a) Electromagnetic energy. (b) Modulation formats.

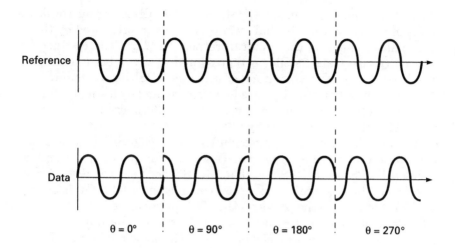

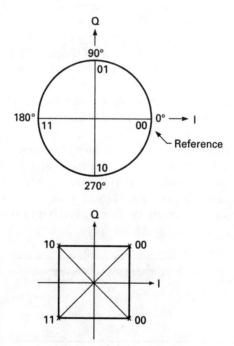

Figure 2.18 QPSK.

The coordinate system for QPSK is best realized if you think in terms of an *XY* coordinate chart where *X* is now represented by the *I*, or in phase, and *Y* is the quadrature portion (*I*,*Q*). The distinct *IQ* location (phase state) is shown represented by a symbol made up of two distinct bits. The advantage of utilizing QPSK is the bandwidth efficiency. Since two data bits are now represented by a single symbol, they require less spectrum to transport the information. The symbol rate = bit rate/number of bits/symbol.

Differential quadrature phase shift keying (DQPSK) is a modulation technique similar to QPSK. However, the primary difference between DQPSK and QPSK is that DQPSK does not require a reference from which to judge the transition. Instead DQPSK's data pattern is referenced to the previous DQPSK's phase state.

DQPSK has four potential phase states with the data symbols defined relative to the previous phase state as shown in Table 2.7.

Pi/4 differential quadrature phase shift keying (Pi/4 DQPSK) modulation is similar to DQPSK. However, the difference between pi/4 DQPSK and DQPSK is that the Pi/4 DQPSK phase transitions are rotated 45° from DQPSK. Like DQPSK, Pi/4 DQPSK has four transition states, defined relative to the previous phase state shown in Table 2.8.

One method utilized to represent phase and amplitude modulation is through *IQ* diagrams. The *IQ* diagram in Fig. 2.19 utilizes vector notation for representing the actual *I* and *Q* values. There are several methods that can be utilized to view digitally modulated signals. Each of the methods has its positive and negative aspects, but the method chosen needs to match the objective at hand. The four primary methods for viewing digitally modulated signals are spectrum display, vector diagram, constellation diagram, and the eye diagram.

TABLE 2.7

Symbol	DQPSK phase transition, degrees
00	0
01	90
10	−90
11	180

TABLE 2.8

Symbol	Pi/4 DQPSK phase transition, degrees
00	45
01	135
10	−45
11	−135

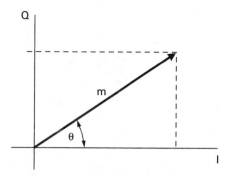

Figure 2.19 *IQ* modulation. [Note: $I = m \cos \theta$, $Q = m \sin \theta$, $m = (I^2+Q^2)^{1/2}$, $\theta = \arctan(Q/I)$.]

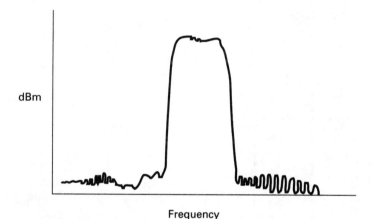

Figure 2.20 Spectrum display.

The spectrum display (Fig. 2.20) is probably the most common method for viewing any RF modulation scheme. However, for viewing digital modulation the spectrum display has limited value for analysis. The primary benefit of utilizing a spectrum display is to view the entire spectrum. The spectrum display can be used to view out-of-band emissions and the sidebands of the digitally modulated signal.

The vector diagram (Fig. 2.21) basically plots the *I* components as a function of *Q*. The primary purpose of utilizing a vector diagram for analysis of a digitally modulated signal is to view the transitions between the various states in a quadrature modulated signal. The transition status can be used to determine the overall modulation quality of the signal being viewed. If there is little error with the signal, the locations on the vector diagram which represent symbol

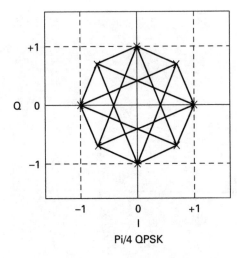

Pi/4 QPSK

Figure 2.21 Vector diagram.

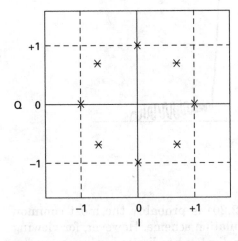

Figure 2.22 Constellation diagram.

points are easily definable and the variation transitions trajectories intersect closely at each of these points.

Another method utilized for viewing digital modulation is the constellation diagram (Fig. 2.22). It shows the relationship between different amplitude and phase states of the modulated signal. The purpose of the constellation diagram is to display the error vector at the symbol sample time. The error vector is the difference between the theoretical symbol location and the actual symbol location on the constellation diagram.

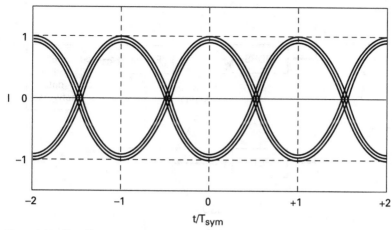

Figure 2.23   Eye diagram.

The eye diagram in Fig. 2.23 is yet another method for the analysis of a digital communication system. For quadrature modulation the eye diagram plots both the I and Q components as a function of time. Utilizing the eye diagram you can determine the phase and amplitude errors of the system. As the number of errors in phase and amplitude increase, the eye pattern closes.

2.9   Cellular Digital Packet Data (CDPD)

CDPD is a packetized data service utilizing its own air interface standard. The CDPD systems utilized by cellular operators are functionally a separate data communication service which physically shares the cell site and cellular spectrum. CDPD has many applications, but most are applicable for short bursty-type data applications and not large file transfers. CDPD applications of short messages are e-mail, telemetry applications, credit card validation, and global positioning, to mention a few potentials.

CDPD does not establish a direct connection between the host and server locations. Instead it relies on the OSI model for packet switching data communications which routes the packet data throughout the network. The CDPD network has various layers which the system comprises; layer 1 is the physical layer, layer 2 is the data link itself, and layer 3 is the network portion of the architecture. CDPD utilizes an open architecture and has incorporated authentication and encryption technology into its airlink standard. The CDPD system consists of several major components. A block diagram of a CDPD system is shown in Fig. 2.24.

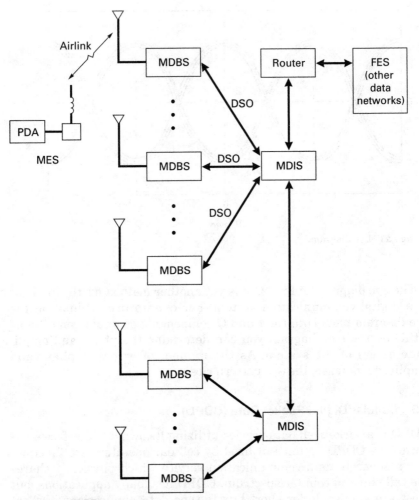

Figure 2.24 CDPD system.

The mobile end system (MES) is a portable wireless computing device that moves around the CDPD network communicating to the MDBS. The MES is typically a laptop computer or other personal data device which has a cellular modem.

The mobile data base station (MDBS) resides in the cell site itself and can utilize some of the same infrastructure that the cellular system does for transmitting and receiving packet data. The MDBS acts as the interface between the MES and the MDIS (mobile data intermediate system). One MDBS can control several physical radio channels depending on the site configuration and loading requirements. The MDBS communicates to the MDIS via a 56-kbits/s data link.

Often the data link between the MDBS and MDIS utilizes the same facilities as the cellular system but occupies a dedicated time slot.

The MDIS performs all the routing functions for CDPD, utilizing the knowledge of where the MES is physically located within the network itself. Several MDISs can be networked together to expand a CDPD network. The MDIS also is connected to a router or gateway which connects the MDIS to a fixed end system (FES). The FES is a communication system that handles layer 4 transport functions and other higher layers.

The CDPD system utilizes a gaussian minimum-shift keying (GMSK) method of modulation and is able to transfer packetized data at a rate of 19.2 kbits/s over the 30-kHz-wide cellular channel. The frequency assignments for CDPD can take on two distinct forms. The first form of frequency assignment is a method of dedicating specific cellular radio channels to be utilized by the CDPD network for delivering the data service. The other method of frequency assignment for CDPD is to utilize channel hopping where the CDPD's MDBS utilizes unused channels for delivering its packets of data. Both methods of frequency assignments have advantages and disadvantages.

Utilizing a dedicated channel assignment for CDPD has the advantage of not having the CDPD system interfere with the cellular system it is sharing the spectrum with. By enabling the CDPD system to operate on its own set of dedicated channels there is no real interaction between the packet data network and the cellular voice network. However, the dedicated channel method reduces the overall capacity of the network, and depending on the system loading conditions this might not be a viable alternative.

If the method of channel hopping is utilized for CDPD, and this is part of the CDPD specification, the MDBS for that cell or sector will utilize idle channels for the transmissions and reception of data packets. In the event the channel that is being used for packet data is assigned by the cellular system for a voice communication call, the CDPD MDBS detects the channel assignment and instructs the mobile end system (MES) to retune to another channel before it interferes with the cellular channel. The MDBS utilizes a scanning receiver or sniffer, which scans all the channels it is programmed to scan to determine which channels are idle or in use.

The disadvantage of the channel hopping method involves the potential interference problem to the cellular system. Coexisting on the same channels with the cellular system can create mobile–to–base station interference. The mobile–to–base station interference occurs because of the different handoff boundaries for CDPD and cellular systems for the same physical channel. The difference in handoff boundaries occurs largely because CDPD utilizes a

BER for handoff determination and the cellular system utilizes receive signal strength indication (RSSI) at either the cell site for the analog or mobile assisted hand-off (MAHO) for the digital system.

2.10   Microcells

Microcells are starting to become prevalent in cellular systems as operators strive to reduce the physical footprint of the macrocell sites. The reduction in the geographic area each cell site covers is meant to facilitate more reuse in the network. Microcells are also deployed to provide coverage in buildings, subway systems, and tunnels, and resolve unique coverage problems. The technology platforms that tend to be referenced as microcells involve any communication system that is less than ½ km in radius.

Currently several types of technology platforms fall into the general categorization of microcells.

1. Fiber-fed microcell
2. T1 microcell
3. Microwave microcell
4. High-power reradiators
5. Low-power reradiators
6. Bidirectional amplifier

The choice of which technology to utilize is driven by a variety of factors unique to that particular situation. One factor is the application being engineered for (capacity, coverage, or private wireless PBX). Another important factor is the configuration options available at that location for providing radio capacity. A third factor is overall cost.

2.10.1   Fiber-fed microcell

The fiber-fed microcell can be and is being deployed in cellular systems. Its choice has many advantages and disadvantages. Currently two distinct types of fiber-fed microcells are available, analog and digital. Both require the use of dark fiber to make them operate in a cellular network. Dark fiber is often not readily available in many of the areas where a microcell could be deployed and requires that the pair of fibers utilized for the microcell be dedicated only to that microcell. However, a fiber-fed microcell can be utilized with any of the cellular infrastructure vendors, making it vendor-transparent.

Figure 2.25 is a drawing of the analog fiber-fed microcell. The positive attributes of this microcell are that it provides easy transport of

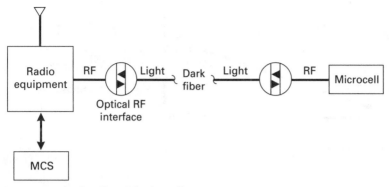

Figure 2.25 Analog fiber-fed microcell.

the RF modulated carrier and various technology platforms. The analog microcell will transport regular FM, TDMA, and CDMA signals because it utilizes AM modulation for transporting the RF information along the dark fiber. The microcells themselves are small in size and are excellent choices for establishing a centralized radio hub site, allowing one cell site to feed many microcells.

The negative attributes of the analog microcell tie into the requirement that it utilize dark fiber, the limited optical link budget, and the expense of the optical interface equipment. The optical interface equipment is needed to convert the RF modulated signal into a light format for transport along the dark fiber. At the other end of the fiber path the optical interface equipment is needed to convert the light format into the RF modulated signal again.

The other type of fiber-fed microcell is the digital fiber-fed microcell. The issues are primarily the same with the digital microcell except it is able to transport the converted RF modulated signal over a greater distance than the analog based unit. The digital microcell, however, may or may not be able to transport TDMA or CDMA signals based on the digitization process utilized. The digital microcell also uses dark fiber to connect itself between the host and donor locations.

2.10.2 T1 microcell

The T1- or copper-fed microcell has many potential applications in the wireless arena. It has the distinct advantage of providing full or near full functionality of a regular cell site in a small package. The T1 microcell generalized configuration is shown in Fig. 2.26. It has the distinct advantage of being able to utilize the readily available T1 facilities in a network. The obvious advantage of the T1 facilities is

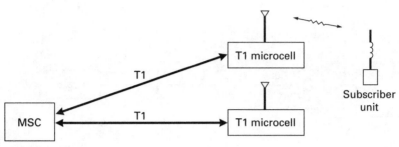

Figure 2.26 T1 microcell.

interconnect ease and lower cost with initial deployment. The reduced footprint of the T1 microcell and its ability to capitalize on the available T1 facilities make it suitable for a variety of applications.

The T1 microcell is presently vendor-specific, however. The T1 microcell is part of cellular vendors' product offering and will not work with other vendors' equipment on its own. It is possible, however, to deploy a T1 microcell system in a network that utilizes another vendor's equipment through the use of intersystem protocols like IS-41.

2.10.3 Microwave microcell

Microwave-fed microcells are another type of microcell. They have some unique advantages associated with this type of technology, which lie in the fact that it is independent of backhaul and infrastructure vendors. This type of microcell is also directly compatible with different technology platforms. A simple microwave microcell system is shown in Fig. 2.27.

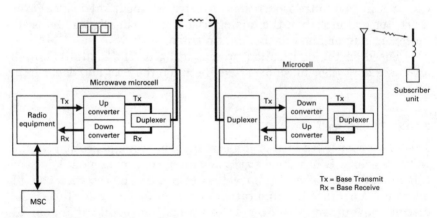

Figure 2.27 Microwave microcell.

The microwave microcell utilizes microwave point-to-point communication to connect the feeding cell site with the remote locations. The advantage with utilizing a point-to-point method for connecting the microcell to the donor cell is the fact it doesn't require the use of landline facilities, T1 or fiber, to connect the donor cell and the microcell. The disadvantages are frequency coordination of the microwave path, ensuring path clearance, and an additional antenna and feedline for the donor cell.

The microwave frequency coordination can place severe operational limitations on the utilization of this technology platform. Specifically the actual microwave path for the frequency of interest might not be available, requiring another operating band. The choice of operating frequency band has a direct impact on the path length, i.e., the distance between the donor cell and the microcell. The choice of operating frequency coupled with the path length determine the path outages the system will experience owing to weather conditions for the region.

The other issue with the microwave microcell is ensuring that there is sufficient path clearance for the path that is utilized. Depending on the microwave path required and the frequency of operation, the donor or microcell's microwave antennas might have to be placed at a significant elevation to ensure a reliable communication path.

2.10.4 Reradiators

A reradiator as the solution to a communication problem or situation has been used effectively by many operators. Reradiators take on three distinct forms, high-power, low-power translating, and non-translating or bidirectional. When utilizing a reradiator the only true objective can be coverage, since this technology platform will not add capacity to the network. The reradiator, or repeater, redistributes capacity from one cell site to another area. It can alter the capacity distribution of the network through the selection of the donor cell from which the capacity is drawn.

The primary negative attribute of a repeater is its ability to add interference into the network as both mobile-to-base and base-to-mobile interference. The interference is a result of the repeater's extending the actual coverage of the donor cell beyond what it was designed to operate. Which type of reradiator to utilize depends on the design objective and the configuration options available for the situation.

High-power reradiators. The high-power reradiator (HP) is one form of reradiator that is available for use in a network. It is an extension of the donor cell utilizing different frequencies and is able to achieve transmit power levels comparable to that of a cell site itself. A simple

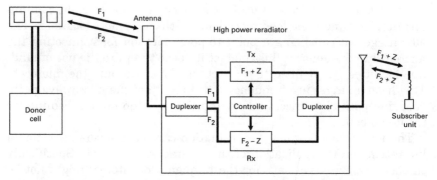

Figure 2.28 High-power reradiator. (Key: F_1 = donor cell Tx frequency, F_2 = donor cell Rx frequency.)

diagram of an HP is shown in Fig. 2.28. The high-power reradiator is often utilized when there is not sufficient isolation between the transmitter and receiver of the reradiator. Isolation is a requirement for low-power reradiators since they transmit the same frequency they receive, in both up- and down-link directions. As with all communication systems that are of a point-to-point nature, it is imperative that a path clearance analysis be performed to ensure proper operation.

The HP is an extension of the donor cell site utilizing different frequencies to communicate with a subscriber unit. The diagram in Fig. 2.28 illustrates that the HP translates the donor cell sites transmit and receive frequencies by some value Z. The translated channel, however, is still within the cellular carrier's operating band and represents a specific cellular channel. The actual channels that HP translates the donor cell channels to are defined by the operator, which requires frequency coordination for interference protection inside the cellular network itself.

The HP also has its own cell site controller which has its own software. The controller software is utilized not only to control the translation of the radio channel frequency but also to perform a call-processing role. The primary call-processing role served by the HP is when the HP tries to hand back the mobile to the system. The controller software instructs the mobile to retune from channel $f_1 + Z$ to just f_1 in an attempt to transfer the mobile to the network for the continuation of the call. The reason the HP tries to hand back the mobile is the limited footprint the HP has in the network.

Low-power reradiators. The low-power reradiator (LP) is similar in configuration and purpose to the HP except that it has lower transmit power. The LP reradiates the signals it receives to improve radio coverage in a given area but does not translate the voice channels to another frequency. The LP, however, does translate the control, or

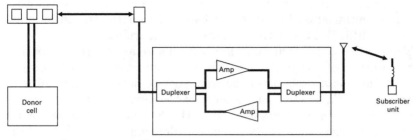

Figure 2.29 Low-power reradiator.

setup, channel to another control channel frequency. Figure 2.29 shows a simplified block diagram of a low-power reradiator.

The LP configuration requires that close attention be paid to the transmit and receive isolation to prevent feedback problems with reradiating the same signal. The minimum isolation usually required for an LP is 70 dB. It is achieved through physical separation of the antennas themselves. However, at times an in-line attenuator is utilized in one branch of the antenna system to obtain the isolation required. The LP, as with all reradiators, requires that the proper path clearance is obtained between the donor and LP unit itself. The primary differences between the HP and LP are the isolation requirements, frequency translations, and output power.

Bidirectional amplifier. The bidirectional amplifier is similar to the low-power reradiator. The block diagram used in Fig. 2.29 to represent a low-power reradiator is functionally the same for a bidirectional amplifier. The bidirectional amplifier, like the low-power reradiator, does not translate voice channel frequency. The bidirectional amplifier, however, does not translate the control channel that it receives from the donor cell site.

The bidirectional amplifier has the same isolation and path clearance requirements as a low-power reradiator. The bidirectional amplifier can be utilized for in-building applications either by itself or as an adjunct to microcell deployments. The bidirectional amplifier offers a cost-effective method of providing coverage in an in-building environment.

2.11 Path Clearance

Radio path clearance is an essential criterion for any point-to-point communication system. There are many different types of point-to-point communication systems that can be utilized in cellular communication. Some of the technology platforms that require a path clearance analysis involve point-to-point microwave and reradiators. The path clearance analysis needs to be performed for every RF point-to-point communication link in the network. The path clearance analysis is not difficult to perform. An example is provided below.

The Fresnel zone is shown in Fig. 2.30. There are effectively an infinite number of Fresnel zones for any communication link. The Fresnel zone is a function of the frequency of operation for the communication link. The primary energy of the propagation wave is contained within the first Fresnel zone. The Fresnel zone is important for the path clearance analysis since it determines the effect of the wave bending on the path above the earth and the reflections caused by the earth's surface itself. The odd-numbered Fresnel zones will reinforce the direct wave and the even Fresnel zones will cancel.

In a point-to-point communication system it is desirable to have at least a 0.6 first Fresnel zone clearance to achieve path attenuation approaching free space loss between the two antennas. The clearance criteria apply to all sides of the radio beam, not just the top and bottom portions represented by the drawing in Fig. 2.30.

Environmental effects on the propagation path have a direct influence on the point-to-point communication system. The environmental effects altering the communication system are foliage, atmospheric moisture, terrain, and antenna height of the transmitter and receiver.

The K factor utilized for point-to-point radio communication is 1.333 or 4/3 earth radius. It ties in the relationship between the earth's curvature and the atmospheric conditions that can bend the electromagnetic wave.

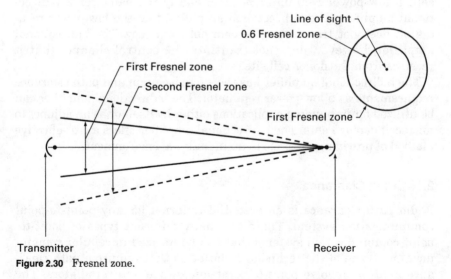

Figure 2.30   Fresnel zone.

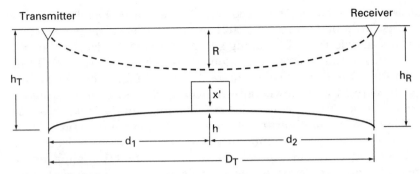

Figure 2.31 Path clearance.

Example To determine the path clearance required for a point-to-point communication site refer to Fig. 2.31.

First Fresnel zone:

$$R = 72 \sqrt{\frac{d_1 d_2}{D_T f}}$$

Earth curvature:

$$h = \frac{d_1 \cdot d_2}{1.5\, k}$$

with f = frequency in GHz, d_1, d_2, D_T in miles; x, h, h_T, h_R in feet, and where d_1 = 1.6 mi, d_2 = 2.1 mi, D_T = 3.7 mi, and f = 0.88 GHz (or 880 MHz).

$$R = 72 \sqrt{\frac{(1.6)\,(2.1)}{(3.7)\,(.88)}} = \sqrt{\frac{3.36}{3.256}} = 73.14 \text{ ft} \qquad R' = (0.6)\,R = 43.884$$

$$h = \frac{(1.6)\,(2.1)}{(1.5)\,(4/3)} = 1.68 \text{ ft} \qquad k = \tfrac{4}{3}$$

Assume that the transmitter, receiver, and obstruction have the same ASML.

Earth curvature	1.68
0.6 Fresnel zone	43.88
Obstruction height	100
	145.56 ft

The minimum T_x and R_x heights for the system are $h_T = h_R = 145.56$ ft.

2.12 In-Building Applications

Microcells have numerous applications for in-building applications. The applications include improving coverage for a convention center or large client, disaster recovery, or a wireless PBX, to mention a few. The propagation of the radio-frequency energy, however, takes on unique characteristics in an in-building application as compared to an

outdoor environment. The primary difference in propagation characteristics for in-building versus outdoor applications is fading, shadowing, and interference. The fading situation results in deeper and spa-
tially closer fades when a system is deployed in an in-building
application. Shadowing is also quite different in an in-building application owing to the lower antenna heights and excessive losses
through floors, walls, and cubicles. The shadowing effects in an in-
building application limit the effective coverage area to almost line of
sight (LOS) for cellular communications. The interference issue with
an in-building system can actually benefit in-building applications
since the interference is primarily noise-driven and not cochannel
interference. In-building systems are primarily noise-driven because
of the attenuation experienced by external cell sites as they trans-
verse into the buildings and various structures.

Some unique considerations must be taken into account regarding
microcell system design for inside a building:

1. Base to mobile power

2. Mobile to base power

3. Link budget

4. Coverage area

5. Antenna system type and placement

6. Frequency planning

The base to mobile power needs to be carefully considered to ensure
that the desired coverage is met, deep fades are mitigated in the area
of concern, the amplifier is not being over- or potentially underdriven,
and mobile overload does not take place. The desired coverage that
the in-building system is to provide might require several transmitters because of the limited output power available from the units
themselves. For example, if the desired coverage area required 1 W
ERP to provide the desired result, a 10-W amplifier would not be able
to perform the task if you needed to deliver a total of 40 channels to
that location, meaning only 25 mW of power per channel was really
available. The power limitation can, and often does, make the limiting path in the communication system for an in-building system the
forward link.

The forward link power problem is further complicated by the fact
that portable and potential mobile units will be operating in very
close proximity to the in-building system's antenna. If the forward
energy is not properly set, a subscriber unit could easily go into gain
compression, causing the radio to be desensitized. The mobile to base
power also needs to be factored into the in-building design. If the

power window (dynamic power control) is not set properly, imbalances could exist in the talk-out to talk-back path. Usually the reverse link in any in-building system is not the limiting factor, but the mobile to base path should be set so that there is a balanced path between the talk-out and talk-back paths.

Most in-building systems have the ability to utilize diversity receive but do not utilize it for a variety of reasons. The primary reason for not utilizing diversity receive in an in-building system is the need to place two distinct antenna systems in the same area.

The link budget for the communication system needs to be calculated in advance to ensure that both the forward and reverse links are set properly. The link budget analysis plays a very important role in determining where to place the antenna system and distributed or leaky feeder, and the number of microcell systems required to meet the coverage area requirement.

The antenna system selected for the in-building application is directly related to the uniformity of the coverage and quality of the system. The antenna system (no diversity) primarily provides LOS coverage to most of the areas desired in the defined coverage area. Based on the link budget requirements the antenna system can be either passive or active. The antenna system for an in-building system may take on the role of having passive and active components in different parts of the system to satisfy the design requirement. Typically a passive antenna system is made up of a single or distributed antenna system or can also utilize a leaky coaxial system. The in-building system shown in Fig. 2.32 utilizes a distributed antenna system for delivering the service. A leaky coaxial system could also be deployed within the same building to provide coverage for the elevator in the building.

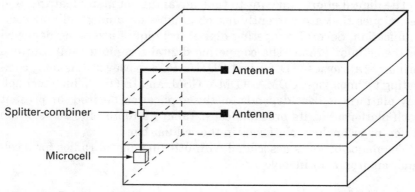

Figure 2.32 Passive in-building system.

The advantage a leaky coaxial system has over a distributed antenna is that it provides a more uniform coverage to the same area. However, the leaky coaxial system does not lend itself to an aesthetic installation in a building. The use of a distributed antenna system for providing coverage in an in-building system makes the communication system stealthy.

If the antenna system requires the use of active devices in the communication path, the level of complexity increases. The complexity increases for active devices since they require ac or dc power and introduce another failure point in the communication system. However, the use of active devices in the in-building system can ultimately make the system work in a more cost-effective fashion. The most common active device in an in-building antenna system is a bidirectional amplifier.

The frequency planning for an indoor microcell system needs to be coordinated with the external cellular network. Most in-building systems are designed to facilitate handoffs between the in-building and external cellular system. If the in-building system is utilizing a microcell with its own dedicated channels assigned to it, it is imperative that the in-building system be integrated into the macrocellular network.

2.13 Digital Cellular Systems

Digital radio technology is currently being deployed in cellular systems with the attempt to increase their quality and capacity. In an analog cellular system the voice communication is digitized within the cell site itself for transport along the T-carrier to the MTSO. The voice representation and information transfer utilized in AMPS is analog. An effort is under way in this part of the communication link to convert to a digital platform.

The digital effort is meant to take advantage of many features and techniques that are currently not obtainable for analog cellular communication. Several competing digital techniques are being deployed in the cellular arena. The competing digital techniques fall into two primary categories, TDMA and CDMA. PCS, however, has four, competing technologies, CDMA, TDMA, GSM, and IS-661. Which technology platform is best depends on the application desired; at present each platform has its pros and cons. Table 2.9 represents some of the different technology platforms in the cellular band.

The major benefits associated with utilizing digital radios for a cellular environment involve

- Increased capacity over analog

TABLE 2.9

	AMPS	NADC	CDMA	GSM*
Frequency range, MHz	824–849	824–849	824–849	890–915
	869–894	869–894	869–894	935–960
Technology	FM	TDMA	CDMA	TDMA
Modulation	FM	pi/4 DQPSK	QPSK	0.3 GMSK
Channel spacing	30 kHz	30 kHz	1.23 MHz	200 kHz
Modulation data rate	NA	48.6 kbits	1.2288 Mbits	270.833 kbits
CODEC	NA	VSELP	(8550 bits/s)	RELP-LTP
		8 kbits		13 kbits
Spectrum allocation	50 MHz	50 MHz	50 MHz	50 MHz
Number of channels	832	832	10	124
Users per channel	1	3	118	8
Standard	AMPS	IS-54	IS-95	GSM

*The GSM technology, in the frequency band listed, is not utilized in the United States.

- Reduced capital infrastructure costs
- Reduced capital per subscriber cost
- Reduced cellular fraud
- Improved features
- Improvement in customer-perceived performance
- Encryption

Figure 2.33 represents the differences between an analog and digital radio. Reviewing the digital radio portion of the diagram, the initial information content, usually voice, is input into the microphone of the transmission section. The speech then is processed in a vocoder which converts the audio information into a data stream utilizing a coding scheme to minimize the number of data bits required to represent the audio. The digitized data then go to a channel coder which takes the vocoder data and encodes the information even more, so it will be possible for the receiver to reconstruct the desired message. The channel coded information is then modulated onto an RF carrier utilizing one of several modulation formats covered previously in this chapter. The modulated RF carrier is then amplified, passes through a filter, and is transmitted out an antenna.

The receiver, at some distance away from the transmitter, receives the modulated RF carrier through use of the antenna, which then passes the information through a filter and into a preamp. The modulated RF carrier is then downconverted in the digital demodulator section of the receiver to an appropriate intermediate frequency. The demodulated information is then sent to a channel decoder, which performs the inverse of the channel coder in the transmitter. The digital information is then sent to a vocoder for voice information recon-

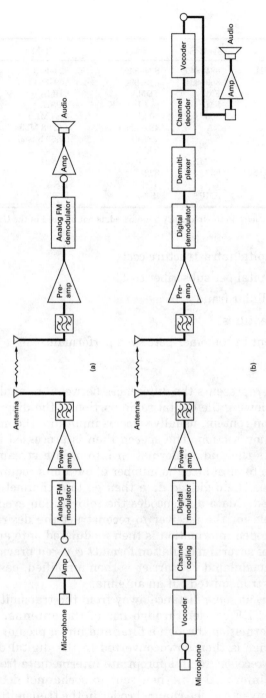

Figure 2.33 (a) Analog and (b) digital radio.

struction. The vocoder converts the digital format into an analog format which is passed to a audio amplifier connected to a speaker for the user at the other end of the communication path to listen to the message sent.

2.14   Radio Path Impairment

The communication quality between a mobile subscriber unit and the cell site depends on a variety of factors affecting the path over which the radio signal travels. Several types of signal impairments take place over the radio signal's path. The four basic impairments experienced in a communication path involve path loss, shadowing, multipath, and doppler shift.

The impact on the communication link from path loss is covered earlier in this chapter. Path loss is a direct result of the distance between the transmitter and receiver in the communication path. Shadowing, also called slow fading, is caused largely by partial blockage or environmental absorption such as trees, and this is covered in an earlier section of this chapter.

2.14.1   Multipath and delay spread

In any communication system multipath propagation presents some of the most challenging problems for designing a mobile communication system. Multipath propagation is the predominant form of transmission path in an urban environment since more than one reflection in the transmission path is normal (Fig. 2.34). The issues of multipath problems to a communication system show up a delay spread and rayleigh fading.

In an analog communication system multipath, the same information taking multiple paths causes fading. The fading itself can sound like a flutter at low speeds. The multipath is a result of rayleigh or fast fading and of the receiving antenna's moving through constructive and destructive wavefronts. The receiver's susceptibility to fading is a function of the frequency of operation and the receiver bandwidth. The higher the frequency the shorter the distance is between wave crests. The wider the bandwidth the less susceptible the receiver is to experience fading.

Delay spread (time dispersion) is to the digital radio system what multipath is to analog. However, with a digital system the delay in the signal's arrival is more important than that of the signal's received level. Dispersion occurs when multiple signals arrive at different times to the receiver and the difference in time between the signals arriving is in the order of a bit period. The multiple signals

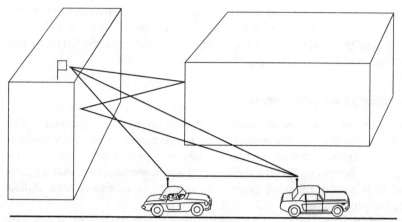

Figure 2.34   Multipath signals in an urban environment.

arriving in the receiver within a bit period cause a distortion in the representation of the bit desired or even the wrong bit's being decoded because the delayed, reflected signal is stronger than the direct signal itself. When the time delay spread is no longer negligible with respect to the modulation bandwidth, the received information will be distorted because of the different paths transferred with the multipath incoming waves. The delay spread is more pronounced with higher data rates since the effect can cause symbols to overlap, producing intersymbol interference. Figure 2.35 is an example of intersymbol interference (ISI). In Fig. 2.35 two multipath signals are received but there is enough delay between the symbols received to cause signal ambiguity, resulting in ISI.

The interesting point about delay spread is that the signal level of the incident, or desired signal, does not play as important a role as does the time delay between received signals. The dispersion problem for a digital communication system could be so severe for a cell site that you could have line of sight (LOS) communication and still not successfully demodulate the signal properly. There are a few solutions that one can utilize for correcting the dispersion problem.

One method of resolving the delay spread or dispersion problem is to employ adaptive equalizers in the demodulation portion of the receiver, utilized for TDMA systems. The equalizer provides a method for detecting the delayed signals and locking onto the strongest signal received. The equalizer operates by utilizing a training sequence which is sent at the start of the data communication burst, and then the equalizer adjusts itself to provide the maximum response on the channel, negating the effects of the radio channel itself. The use of equalizers is an integral part of a TDMA cellular communication.

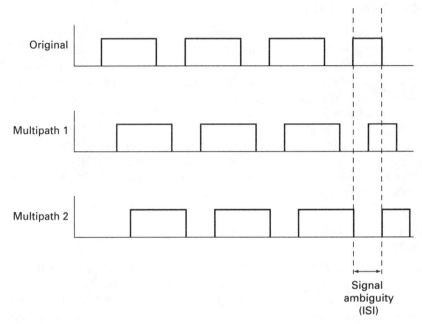

Figure 2.35 ISI.

A CDMA system, however, does not utilize an adaptive equalizer to try to mitigate the effect of delay spread. Instead the CDMA system utilizes a rake receiver which has the ability to discriminate, or pick out, the desired signal from the other. The rake receiver is an integral part of a CDMA communication system.

One additional method that can be utilized in minimizing dispersion problems is through reducing the effective radiated power of the offending cell site itself. However, reducing the ERP of a site will also have a negative impact on the effective coverage area and may require more cell sites to provide the same level of coverage. Another method for reducing dispersion problems is to alter the antenna orientation of the cell site, therefore changing the dispersion path which is negatively affecting the radio communication.

2.14.2 Doppler shift

When there is any motion between the transmitter and receiver a frequency (doppler) shift is experienced between the transmitted and received frequency. Depending on the relative change, the doppler shift can be serious enough to produce a random frequency modulation on the original signal itself. It also can have a positive or negative impact on the multipath propagation.

2.15   Spread Spectrum

Spread spectrum is used to describe a variety of different technology platforms in wireless communications. A spread spectrum system is any communication system that deliberately occupies more channel bandwidth than the minimum required for data transfer. The rationale behind utilizing spread spectrum is to gain an improvement in the signal-to-noise ratio of the communication system itself. Three basic types of spread spectrum formats are utilized with many perturbations. The three basic spread spectrum types are code-division multiple access (CDMA), time-division multiple access (TDMA), and frequency-division multiple access (FDMA). An example of an FDMA system in wireless is a cellular system.

2.15.1   Code-division multiple access (CDMA)

Code-division multiple access (CDMA) is a spread spectrum technology platform which enables multiple users to occupy the same radio channel (frequency spectrum) at the same time. CDMA has been and is being utilized for microwave point-to-point communication, satellite communication, and also by the military. With CDMA each of the subscribers, or users, utilizes a unique code to differentiate themselves from the other users. CDMA offers many unique features, including the ability to thwart interference and improved immunity to multipath effects due to its bandwidth. The benefits associated with CDMA are:

- Increased system capacity over analog and TDMA
- Improved interference protection
- No frequency planning required between CDMA channels
- Improved handoffs with MAHO and soft handoffs
- Fraud protection due to encryption and authentication
- Accommodation of new wireless features

CDMA is based on the principle of direct sequence (DS) and is a wideband spread spectrum technology. The CDMA channel utilized is reused in every cell of the system and is differentiated by the pseudorandom number (PN) code that it utilizes.

Guard band and guard zone.   The introduction of CDMA into an existing AMPS system will require the establishment of a guard band and guard zone. The guard band and guard zone are required for CDMA to ensure that the interference received from the AMPS system does not prevent CDMA from performing well.

The specific location that the CDMA channel or channels occupy in a cellular system is dependent upon a multitude of issues. The first issue is how much spectrum will be dedicated to the use of CDMA for the network. The spectrum issue ties into the fact that one CDMA channel occupies 1.77 MHz of spectrum, 1.23 MHz per CDMA channel, and 0.27 MHz of guard band on each side of the CDMA channel. With a total of 1.77 MHz per CDMA the physical location in the operator's band that CDMA will operate in needs to be defined. Presently for the B-band carrier (wireline operators) two predominant locations are being utilized. The first location in the spectrum is the band next to the control channels and the other section is in the lower portion of the extended AMPS band. The upper end of the EAMPS band is not as viable owing to the potential of AGT interference since AGT transmit frequencies have no guard band between AMPS receive and AGT transmit. The lower portion of the EAMPS band has the disadvantage of receiving A-band mobile-to-base interference, which will limit the size of the CDMA cell site. Sharper filters could be utilized for the lower EAMPS band, but group delay problems with the use of high selective filters then become a problem.

The other issue with the guard band ties into the actual amount of spectrum that will be unavailable for use by AMPS subscribers in the cellular market. With the expansive growth of cellular systems, assigning 1.77 MHz of spectrum to CDMA reduces the spectrum available for AMPS usage by 15 percent or 59 radio channels from the channel assignment chart. The reduction in the available number of channels for regular AMPS requires the addition of more cell sites to compensate for the radio channels no longer available for use in the AMPS system. Utilizing a linear evaluation the reduction in usable spectrum by 15 percent involves a reduction in traffic-handling capacity by the AMPS system by a maximum of 21 percent at an Erlang B 2 percent GOS with a maximum of 16 channels per sector versus 19. The reduction of 21 percent in initial AMPS traffic-handling capacity requires building of more analog cell sites to compensate for this reduction in traffic-handling ability. The only way to offset the reduction in traffic-handling capacity experienced by partitioning the spectrum is to preload the CDMA subscriber utilizing dual-mode phones or to build more analog cell sites, or to do a combination of both.

The guard zone is the physical area outside the CDMA coverage area that can no longer utilize the AMPS channels now occupied by the CDMA system. Figure 2.36 shows a guard zone versus a CDMA system coverage area. The establishment and size of the guard zone are dependent upon the traffic load expected by the CDMA system.

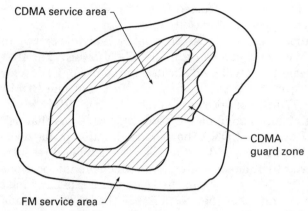

CDMA service area

CDMA
guard zone

FM service area

Figure 2.36   Guard band and guard zone.

The guard zone is usually defined in terms of a signal strength level from which analog cell sites operating with the CDMA channel sets cannot contribute to the overall interference level of the system. The guard zone becomes interesting when the operator on one system wishes to utilize CDMA and requires the adjacent system operator to reduce channel utilization to accommodate the neighbor's introduction of this new technology platform.

For the CDMA system deployed, two distinct methods of implementation are to be considered. The first method is to deploy CDMA in every cell site, for the defined service area, on a one-on-one basis. The other method is to deploy CDMA on an N-to-one basis. Both strategies have advantages and disadvantages (Table 2.10). The deployment strategies are shown in Figs. 2.37 and 2.38. A third potential deployment strategy combines the one-to-one and N-to-one methods.

CDMA spreads the energy of the RF carrier as a direct function of the chip rate that the system operates at. The CDMA system utilizing the Qualcom technology has a chip rate of 1.228 MHz. The chip rate is the rate at which the initial data stream (original information) is

TABLE 2.10

	Advantages	Disadvantages
One to one	Consistent coverage	Cost
	Facilitates gradual growth	Guard zone requirements
	Integrates into existing system	Digital to analog boundary handoff
	Easier to engineer	Slower deployment
	Larger initial capacity gain	
N to one	Lower cost	Harder to engineer properly
	Faster to implement	Lower capacity gain

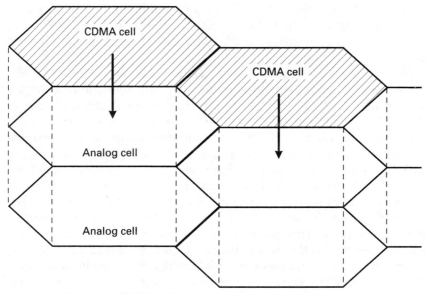

Figure 2.37   One-to-one CDMA analog deployment.

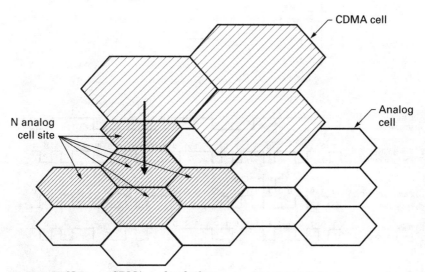

Figure 2.38   N-to-one CDMA analog deployment.

encoded and then modulated. The chip rate is the data rate output of the PN generator of the CDMA system. A chip is simply a portion of the initial data or message which is encoded through use of an XOR process.

The receiving system also must despread the signal utilizing the exact same PN code sent through an XOR gate that the transmitter utilized in order to properly decode the initial signal. If the PN generator utilized by the receiver is different or is not in synchronization with the transmitter's PN generator, the information being transmitted will never be properly received and will be unintelligible. Figure 2.39 represents a series of data that is encoded, transmitted, and then decoded back to the original data stream for the receiver to utilize.

The chip rate also has a direct effect on the spreading of the CDMA signal. Figure 2.40 shows a brief summary of the effects of spreading the original signal that the chip rate chosen has on the original signal. The heart of CDMA lies in the point that the spreading of the initial information distributes the initial energy over a wide bandwidth. At the receiver the signal is despread through reversing the initial

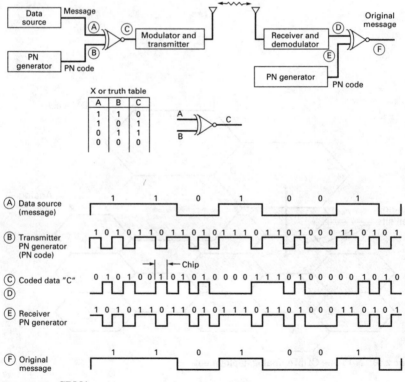

Figure 2.39 CDMA.

Using PN sequence and transmitter with chip (PN) duration of T/L

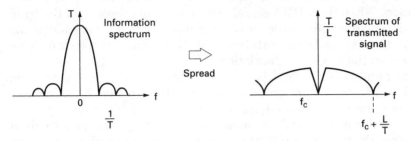

Using correlation and a synchronized replica of the PN sequence at the receiver

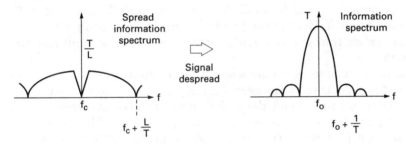

When interference is present

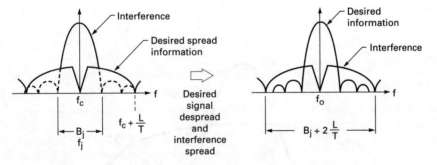

Figure 2.40 Summary of spread spectrum. (Key T/L = chip duration, f_j = jamming frequency, B_j = jammer's bandwidth.)

spreading process where the original signal is reconstructed for utilization. When the CDMA signal experiences interference in the band the despreading process despreads the initial signal for use but at the same time spreads the interference so it minimizes its negative impact on the received information.

The number of PN chips per data bit is referred to as the processing gain and is best represented by the equation in Fig. 2.41. Another way of referencing processing gain is the amount of jammer, interference, and power that is reduced going through the despreading process. Processor gain is the improvement in the signal-to-noise ratio of a spread spectrum system.

Handoff.   The handoff process for CDMA can take on several variants. The variants for handoffs in CDMA are soft handoff, softer handoff, and hard handoff. Each of the handoff scenarios is a result of the particular system configuration and of where the subscriber unit is in the network.

The handoff process begins when a mobile unit detects a pilot signal that is significantly stronger than any of the forward traffic channels assigned to it. When the mobile unit detects the stronger pilot channel, the following sequence should take place. The subscriber unit sends a pilot strength measurement message to the base station instructing it to initiate the handoff process. The cell site then sends a handoff direction message to the mobile unit, directing it to perform

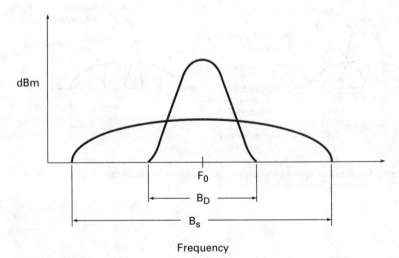

Figure 2.41 Processor gain. (Key: Processor gain $= G_p = B_S/B_D$, B_D = bandwidth of initial signal, B_S = bandwidth of initial signal spread.)

the handoff. Upon the execution of the handoff direction message the mobile unit sends a handoff completion message on the new reverse traffic channel.

In CDMA a soft handoff involves an intercell handoff and is a make-before-break connection. The connection between the subscriber unit and the cell site is maintained by several cell sites during the process. Soft handoff can occur only when the old and new cell sites are operating on the same CDMA frequency channel.

The advantage of the soft handoff is path diversity for the forward and reverse traffic channels. Diversity on the reverse traffic channel results in less power being required by the mobile unit, reducing the overall interference, which increases the traffic-handling capacity.

The CDMA softer handoff is an intracell handoff occurring between sectors of a cell site and is a make-before-break type. The softer handoff occurs only at the serving cell site.

The hard handoff process is meant to enable a subscriber unit to hand off from a CDMA call to an analog call. The process is functionally a break-before-make and is implemented in areas where there is no longer CDMA service for the subscriber to utilize while on a current call. The continuity of the radio link is not maintained during the hard handoff. A hard handoff can also occur between two distinct CDMA channels which are operating on different frequencies.

Forward CDMA channel.   The forward CDMA channel (Fig. 2.42) consists of the pilot channel, one sync channel, up to seven paging channels, and potentially 563 traffic channels. The cell site transmits the pilot and sync channels for the mobile unit to use when acquiring and synchronizing with the CDMA system. When this occurs, the mobile unit is in the mobile station initiation state. The paging channel also transmitted by the cell site is used by the subscriber unit to monitor and receive messages that might be sent to it during the mobile station idle state or system access state.

The pilot channel is continuously transmitted by the cell site. Each cell site utilizes a time offset for the pilot channel to uniquely identify the forward CDMA channel to the mobile unit. There are a possible 512 different time offset values for the cell site to utilize. If multiple CDMA channels are assigned to a cell site the cell will still utilize only one time offset value. The time offset is utilized during the handoff process.

The sync channel is a forward channel that is used during the system acquisition phase. Once the mobile unit acquires the system, it will not normally reuse the sync channel until it powers on again. The sync channel provides the mobile unit with the timing and system

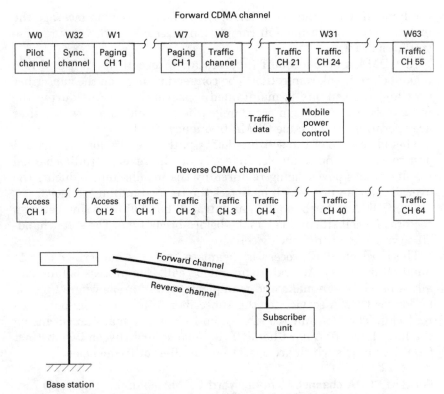

Figure 2.42   CDMA channels. (Key: W = Walsh codes.)

configuration information. The synch channel utilizes the same spreading code (time offset) as the pilot channel for the same cell site. The sync channel frame is the same length as the pilot PN sequence. The information sent on the sync channel is the paging channel rate and the time of the base station's pilot PN sequence with respect to the system time.

The cell site utilizes the paging channel to send overhead information and subscriber specific information. It will transmit at the minimum one paging channel for each supported CDMA channel that has a sync channel.

Once the mobile unit has obtained the paging information from the sync channel the mobile unit will adjust its timing and begin monitoring the paging channel. Each mobile unit, however, monitors only a single paging channel. The paging channel conveys four basic types of information. The first set of information conveyed by the paging channel is the overhead information. The overhead information conveys the system's configuration by sending the system and access parameter messages, the neighbor lists, and CDMA channel list messages. Paging is another message type sent where a mobile unit is paged by the cell site for a land-to-mobile or mobile-to-mobile call. The channel assignment's messages allow the base stations to assign a mobile unit to the traffic channel, alter the paging channel assignment, or redirect the mobile unit to utilize the analog FM system. The forward traffic channel is used for the transmission of primary or signaling traffic to a specific subscriber unit during the call. It also transmits the power control information on a subchannel continuously as part of the closed-loop system. The forward traffic channel will also support the transmission of information at 9600, 4800, or 1200 bits/s utilizing a variable rate which is selected on a frame-by-frame basis, but the modulation symbol rate remains constant.

Reverse CDMA channel. The cell site contiguously monitors the reverse access channel to receive any message that the subscriber unit might send to the cell site during the system access state. The reverse CDMA channel consists of an access channel and the traffic channel. The access channel provides communication from the mobile unit to the cell site when the subscriber unit is not utilizing a traffic channel. One access channel is paired with a paging channel and each access channel has its own PN code. The mobile unit responds to the cell sites messages sent on the paging channel by utilizing the access channel.

The forward and reverse control channels utilize a similar control structure which can vary from 9600, 4800, 2400, or 1200 bits/s, which enables the cell or mobile unit to alter the channel rate dynamically to adjust for the speaker. When there is a pause in the speech, the channel rate decreases so as to reduce the amount of energy received by the CDMA system, increasing the overall system capacity.

There are four basic types of control messages on the traffic channel, those that will control the call itself, handoff messages, power control, security, and authentication. CDMA power control is fundamentally different from that utilized for AMPS or IS-54. The primary difference is that the proper control of total power coming into the cell site, if limited properly, will increase the traffic-handling capability of that cell site. As more energy is received by the cell site its traffic-handling capabilities will be reduced unless it is able to reduce the power coming into it.

The forward traffic power control is composed of two distinct parts. The first part is the cell site which will estimate the forward links transmission loss utilizing the mobile subscribers' received power during the access process. Based on the estimated forward link path loss the cell site will adjust the initial digital gain for each of the traffic channels. The second part of the power control involves the cell site, making periodic adjustments to the digital gain, which is done in concert with the subscriber unit.

The reverse traffic channel signals arriving at the cell site vary significantly and require a different algorithm to be used than that of the forward traffic power control. The reverse channel also has two distinct elements used for making power adjustments. The two elements are the open-loop estimate of the transmit power which is performed solely by the subscriber unit without any feedback from the cell site itself. The second element is the closed-loop correction for these errors in the estimation of the transmit power. The power control subchannel is continuously transmitted on the forward traffic channel every 1.25 ms, instructing the mobile unit to either power up or power down, affecting its mean power output level. There are a total of 16 different power control positions.

2.15.2 Time-division multiple access (TDMA)

Time-division multiple access (TDMA) is another form of spread spectrum technology allowing multiple users to occupy the same frequency spectrum. TDMA technology allows multiple users to occupy the same channel through the use of time division. The TDMA format utilized in the United States follows the IS-54 standard and is referred to as the North American dual mode cellular (NADC) format.

TDMA, utilizing the IS-54 standard, is currently deployed by several cellular operators in the United States. IS-54 utilizes the same channel bandwidth as does analog cellular, 30 kHz per physical radio channel. However, TDMA enables three and possibly six users to operate on the same physical radio channel at the same time. The TDMA channel presents a total of six time slots in the forward and reverse direction. TDMA at present utilizes two time slots per subscriber with the potential to go to half-rate vocoders which require the use of only one time slot per subscriber. TDMA has many advantages in its deployment in a cellular system:

- Increased system capacity up to three times over analog
- Improved protection for adjacent channel interference
- Authentication
- Voice privacy
- Reduced infrastructure capital to deploy
- Frequency plan integration over CDMA

- Short message paging

Integrating TDMA into an existing cellular system can be done more easily than for the deployment of CDMA. The use of TDMA in a network requires the use of a guard band to protect the analog system from the TDMA system. However, the guard band required consists of only a single channel on either side of the spectrum block allocated for TDMA use. Depending on the actual location of the TDMA channels in the operator's spectrum it is possible to require only one or no guard band channel.

The TDMA (IS-54) has the unique advantage of affording the implementation of digital technology into a network without elaborate engineering requirements. The implementation advantages mentioned for TDMA also facilitate the rapid deployment of this technology into an existing network. The implementation of TDMA is further augmented by requiring only one channel per frequency group as part of the initial system offering. The advantage with requiring only one channel per sector in the initial deployment is the minimization of capacity reduction for the existing analog network. Another advantage with deploying one TDMA channel per sector initially eliminates the need to preload the subscriber base with dual-mode (TDMA) handsets.

The TDMA system signaling is shown in Fig. 2.43. The format of the TDMA signal involves a total of six potential conversations taking place over the same bandwidth that one 30-kHz voice conversation utilizes for an analog cellular system. Presently full-rate vocoders are utilized by operators. The full-rate vocoder utilizes two time slots in both the forward and reverse links. The use of the full-rate vocoder puts the limit at three to the number of TDMA users that can occupy one physical channel. In Fig. 2.43 a subscriber utilizing a full-rate vocoder would occupy two time slots, A1 and A2, while a half-rate vocoder user would use only A1 or A2. The time slots are paired in TDMA for a full-rate vocoder system. Conversation A would utilize time slots 1 and 4, conversation B would use slots 2 and 5, and conversation C would utilize time slots 3 and 6.

The modulation scheme utilized by the NADC TDMA system is a pi/4 DQPSK format. The C/I levels used for frequency management associated with TDMA are the same for analog, 17 dB C/I. Access to the TDMA system is achieved through either the primary control channel, utilized for analog communication, or the secondary dedicated control channel. During the initial acquisition phase the mobile reads the overhead control message from the primary control channel and determines if the system is digital capable. If it is, a decision is made whether to utilize the primary or secondary dedicated control channel. The secondary dedicated control channels are assigned as

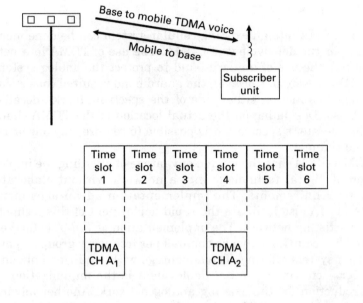

Time slot 1	Time slot 2	Time slot 3	Time slot 4	Time slot 5	Time slot 6

TDMA CH A₁

TDMA CH A₂

Sync.	SACCH	Data	CDVCC	Data	Rsvd.

(Base to mobile)

G	R	Data	Sync.	Data	SACCH	CDVCC	Data

(Mobile to base)

Sync   = synchronization and training channel   G   = guard time
SACCH = slow associated control channel   R   = ramp time
Data   = speech info or FACCH   FACCH = fast associated
CDVCC = coded digital verification color code   control channel
Rsvd. = reserved

Figure 2.43 TDMA channels.

FCC channels 696 to 716 for the A-band system and channels 717 through 737 for the B-band system. The use of the secondary dedicated control channels enables a variety of enhanced features to be provided by the system operator to the subscribers.

One unique feature associated with TDMA is the ability for a mobile assisted handoff (MAHO). The MAHO process enables the mobile unit to constantly report back to the cell site indicating its present condition in the network. The cell site also is collecting data on the mobile unit through the reverse link measurements but the forward link, base to mobile, is being evaluated by the mobile itself, therefore providing critical information about the status of the call.

For the MAHO process the mobile unit measures the received signal strength level (RSSI) received from the cell site. The mobile unit also performs a bit error rate test (BER) and a frame error rate test (FER) as another performance metric. It also measures the signals from a maximum of six potential digital handoff candidates utilizing either a dedicated control channel or a beacon channel. The channels utilized by the mobile unit for the MAHO process are provided by the serving cell site for the call. The dedicated control channel is either the primary or secondary control channel, and the measurements are performed on the forward link. The mobile can also utilize a beacon channel for the performance measurement. The beacon channel is either a TDMA voice channel or an analog channel, both of which are transmitting continuously with no dynamic power control on the forward link. The beacon channel is utilized when the setup or control channel for the cell site has an omni configuration and not a dedicated setup channel per sector.

References

1. Aidarous, Plevyak, *Telecommunications Network Management into the 21st Century,* IEEE Press, New York, 1993.
2. American Radio Relay League, *The ARRL 1986 Handbook,* 63d ed., The American Radio Relay League, Newington, Conn., 1986.
3. *The ARRL Antenna Book,* 14th ed., The American Radio Relay League, Newington, Conn., 1984.
4. AT&T, *Engineering and Operations in the Bell System,* 2d ed., AT&T Bell Laboratories, Murry Hill, N.J., 1983.
5. Boucher, *The Cellular Radio Handbook,* Quantum Publishing, Mendocino, Calif., 1990.
6. Brewster, *Telecommunications Technology,* Wiley, New York, 1986.
7. Bullington, K., "Radio Propagation for Vehicular Communications," *IEEE Transactions on Vehicle Technology,* vol. VT-26 (4): 295–308, 1977.
8. Calhoun, *Digital Cellular Radio,* Artech House, Inc., Norwood, Mass., 1988.
9. Carlson, A. B., *Communication Systems,* 2d ed., McGraw-Hill, New York, 1975.
10. Carr, J. J., *Practical Antenna Handbook,* Tab Books, McGraw-Hill, Blue Ridge Summit, Pa., 1989.
11. Chorafas, *Telephony: Today and Tomorrow,* Prentice-Hall, Englewood Cliffs, N.J., 1984.
12. Code of Federal Regulations, CFR 47 Parts 1, 17, 22, and 24.
13. DeGarmo, Canada, Sullivan, *Engineering Economy,* 6th ed., Macmillan, New York, 1979.
14. DeRose, *The Wireless Data Handbook,* Quantum Publishing, Inc., Mendocino, Calif., 1994.
15. Dixon, *Spread Spectrum Systems,* 2d ed., Wiley, New York, 1984.
16. EIA/TIA Interim Standard, Cellular System Dual-Mode Mobile Stations-Base Station Compatibility Standard, EIA/TIA/IS-54-B, Electronic Industries Associates, Washington, D.C., April 1992.
17. Fike, Friend, *Understanding Telephone Electronics,* 7th ed., Howard W. Sams & Co., 1987.

18. Hata, M., "Empirical Formula for Propagation Loss in Land Mobile Radio Services," *IEEE Transactions on Vehicle Technology*, vol. VT-29 (3): 317–325, 1980.
19. Hess, *Land-Mobile Radio System Engineering*, Artech House, Norwood, Mass., 1993.
20. IEEE C95.1-1991, IEEE Standard for Safety Levels with Respect to Human Exposure to Radio Frequency Electromagnetic Fields, 3 kHz to 300 GHz.
21. *IEEE Standard Test Procedures for Antennas*, IEEE Press, New York, 1979.
22. ITT, *Reference Data for Radio Engineers*, 6th ed., Howard W. Sams & Co., New York, 1983.
23. Jakes, W. C., *Microwave Mobile Communications*, IEEE Press, New York, 1974.
24. Johnson, R. C., and H. Jasik, *Antenna Engineering Handbook*, 2d ed., McGraw-Hill, New York, 1984.
25. Jordan, Balmain, *Electromagnetic Waves and Radiating Systems*, 2d ed., Prentice-Hall, Englewood Cliffs, N.J., 1968.
26. Kaufman, M., and A. H. Seidman, *Handbook of Electronics Calculations*, 2d ed., McGraw-Hill, New York, 1988.
27. Keiser, Strange, *Digital Telephony and Network Integration*, Van Nostrand Reinhold, Princeton, N.J., 1985.
28. Keller, Warrack, Bartel, *Statistics for Management and Economics: A Systematic Approach*, Wadsworth Pub., Belmont, Calif., 1987.
29. Lathi, *Modern Digital and Analog Communication Systems*, CBS College Printing, New York, 1983.
30. Lee, F. E., *Telephone Theory, Principles and Practice*, abc Teletraining Inc., Geneva, Ill., 1985.
31. Lee, W. C. Y., *Mobile Cellular Telecommunications Systems*, McGraw-Hill, New York, 1989.
32. Lee, W. C., *CoChannel Interference Reduction by Using a Notch in Tilted Antenna Pattern*, IEEE, New York, 1985.
33. Lee, W. C., *Effects on Correlation between Two Mobile Radio Base-Station Antennas*, IEEE, New York, 1972.
34. Lindenburg, *Engineering Economics Analysis*, Professional Publications, Inc., 1993.
35. "Low K Comparison of Urban Propagation Models with CW-Measurements," 42nd IEEE VT Conference, Denver, May 1992, pp. 317–325.
36. MacDonald, "The Cellular Concept," *Bell System Technical Journal*, vol. 58, no. 1, 1979.
37. Meyer, *Data Communications Practice*, abc TeleTraining, Inc., Geneva, Ill., 1979.
38. Okumura, Y., E. Ohmori, T. Kawano, and K. Fukuda, "Field Strength and Its Variability in VHF and UHF Land-Mobile Radio Service," *Review of the ECL*, vol. 16, pp. 825–873, 1968.
39. Qualcom, "An Overview of the Application of Code Division Multiple Access (CDMA) to Digital Cellular Systems and Personal Cellular Networks," Qualcom, San Diego, Calif., May 21, 1992.
40. Skolnik, M. I., *Introduction to Radar Systems*, 2d ed., McGraw-Hill, New York, 1980.
41. Simo, "IS-95 Based SS-CDMA: Operational Issues," seminar, January, 1995.
42. Stimson, *Introduction to Airborne Radar*, Hughes Aircraft Company, El Segundo, Calif., 1983.
43. White, Duff, *Electromagnetic Interference and Compatibility*, Interference Control Technologies, Inc., Gainesville, Ga., 1972.
44. Yarbrough, *Electrical Engineering Reference Manual*, 5th ed., Professional Publications, Inc., Belmont, Calif., 1990.

3

Basic Network Components and Key Concepts

Any person involved in the management, design, or operation of a cellular system needs to understand some of the basic components and concepts that make up a wireless network. This chapter provides some description of these basic network building blocks for use in understanding the material presented later in this book as well as expanding the reader's background in the telecommunications field.

3.1 Basic Switching Concepts

The role of the switch in the network has grown along with the importance and complexity of the network itself. This progression began with first the manual exchange (also referred to as just a switch), followed by the rotary exchange and the crossbar exchange, and eventually led up to the development of the modern electronic stored program control (SPC) exchange. Even among the newer switches there are various designs and functional applications depending upon who the switch manufacturer is and what specific role the switch serves in the network, i.e., central office, tandem, private branch exchange (PBX), etc. Regardless of type, all these switches have the same basic function: to route call traffic and to concentrate subscriber line traffic. Some of the more common switch concepts and designs are briefly described below along with a few example network applications.

3.1.1 Switching functions

The numerous functions of the switch within the network can be categorized into three basic groups: elementary functions, advanced func-

TABLE 3.1   Switching Functions

Elementary switching functions:

- Interconnection of input and output line circuits
- Control of communication traffic across line circuit groups

Advanced switching functions:

- Digit analysis
- Call record generation
- Route selection
- Fault detection

Intermediate switching functions:

- Monitor subscriber line circuits

tions, and intermediate functions (see Table 3.1). Elementary switch functions include the process of connecting individual input and output line circuits (trunks) within the switch itself and the ability to control the distribution of communication traffic across clustered groups of line circuits (trunk groups). The ability to interconnect individual line circuits allows the transfer of voice or data signals between end subscriber units or between network nodes to take place in a controlled and selective manner. The switch's ability to direct or route traffic between individual line circuits, based upon larger defined groups of line circuits, allows for more efficient and reliable control of large volumes of system traffic.

The more advanced functions are digit analysis, generation of call records, route selection, and fault detection. Digit analysis is the process of receiving the digits dialed by the subscriber, analyzing them, and determining what action the switch should perform based upon this information, i.e., attempt to place a call to another subscriber, connect the subscriber to an operator for calling assistance, provide a recorded announcement stating that the digits dialed were in error, etc. The process of generating a record or multiple records for any calling activity taking place within the switch is crucial to creating the corresponding billing records for these calls, which in the end process results in the final bill's being completed for the customer. It is important therefore that the switch produce an accurate account (record) of all call processing activity it performs. The route selection function directs all system traffic within the switch to a specific set of facilities (transmission circuit, service circuit, etc.) based upon routing tables developed and maintained by the equipment vendor and the system operator. Finally, the detection of errors or problems occurring within the switch's own hardware and software plus the identification of failures with any of the interconnected facilities is a required function of the switch to ensure the operating quality of the network.

An example of an intermediate switch function is the monitoring of subscriber lines. This function involves the completion of regularly scheduled checks of all line circuits interconnected to the switch for proper operation, i.e., if the circuit is still functional and able to carry system traffic. If a circuit is not performing properly, it will be taken out of service and the fault-detection function will be notified to alert the operations staff.

3.1.2 Circuit switching

Telephone networks utilize circuit switches for the processing and routing of subscriber calls. Circuit switching can be the space-division type, the time-division type, or a combination of these two designs. These designs are utilized in the matrix of the circuit switch. The matrix is where the actual switching of line circuits or trunks takes place.

Space-division switching.   In space-division switches the message paths are separated by space within the matrix, as indicated by the name. In Fig. 3.1 a simple space-division matrix is shown. At each end of the matrix are the wires (actual subscriber lines, line circuits, trunks, etc.) available for switching. This is represented as 1 to N input lines and 1 to N output lines. The input j is connected to output k by closing the crosspoint (relay, contact, semiconductor gate, etc.)

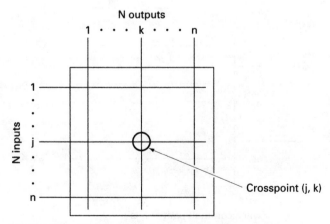

Figure 3.1   An $N \times N$ space-division multiplexing example.

(j, k). The concept here is that only one row of input can be connected to a single column of output. In networks utilizing space switches, each call has its own physical path through the network.

Time-division switching. In time-division switches the message paths are separated in time. Again, the name is derived. The diagram of a time-division switch in Fig. 3.2 will serve as a simple means to explain this switching concept. The subscriber units J_1 through J_n are in conversation with subscriber units K_1 through K_n by means of a time-division switch. The actual input (originating subscriber) and output (terminating subscriber) line circuits are opened and closed by individual switching devices and indicated as A_1 through A_n and B_1 through B_n. (The use of the input and output line circuits does not matter in this explanation. It shows some similarity to the space-division switch matrix example.) In the time-division matrix the connection of subscribers takes place by controlling the operation of the selected switching devices A_1-A_n and B_1-B_n. For example, to connect subscriber J_1 to subscriber K_3, devices A_1 and B_3 must close at the specified time [or time slot, since it is a duration (slot) of time that the connections are allowed to stay closed], and a sample of the speech signal of the conversation between J_1 and K_3 is transmitted over the common path or switch bus. This path or bus is referred to as common because each subscriber conversation will eventually use it at various points during the call. Subscribers J_2 and K_2 may be allowed

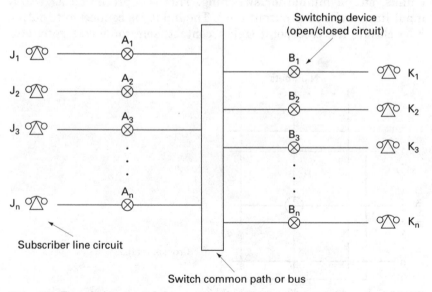

Figure 3.2 Time-division switch.

to converse by closing devices A_2 and B_2 during another time slot. The opening and closing of the switching devices is completed according to the schedule of the n time slots. For instance, if subscribers J_1 and K_3 are in conversation and their sampled speech data are transmitted to one another at time slot 5, then once that time slot has expired the data transmission between these two subscribers will cease and time slot 6 will begin along with the transmission of two different subscribers. No more data from the conversation between J_1 and K_3 will be transmitted until time slot 5 is rescheduled for processing (interconnecting of the designated line circuits to the switch common path or bus). Time slot 5 is scheduled to occur after every n time slots have transpired.

For digital time-division switches to operate, the incoming transmitted voice signals for every phone call must be in a digitized and encoded format. A T-carrier transmission system can provide the proper format to allow direct interconnection to a digital time-division switch without any additional conversion equipment. A description of a T-carrier system is given in Sec. 3.2 along with a description of the pulse-code modulation (PCM) and time-division multiplex (TDM) processes that are used in the T-carrier system. In summary, space-division switching involves the switching of actual circuit interconnections while time-division switching involves the switching of actual digitized voice samples within the switch's matrix.

3.1.3 Packet switching

Voice communication systems utilize circuit switches to provide the switching function of voice line circuits. In data communication systems packet switches are used to perform the switching of data packets between the various nodes and computers in the network. Unlike the longer-duration calls that require circuit switching in a telephone network, packet switching is better suited for the short-burst-like transmissions of the data network. Packet switching involves the sorting of data packets from a single line circuit and switching them to other circuits within the network. These sorting and switching functions are based upon an embedded network address within the data packet itself. An example of a packet switch is a signal transfer point (STP) in an SS7 data network. A description of an STP and corresponding SS7 network is given in Sec. 3.3.

3.1.4 Circuit switching applications

Within the North American public switched telephone network (PSTN), sometimes referred to as the "landline telephone system," there are five classes of switches. These classes can be divided into the central office class 5 level switch and the remaining tandem-type switches of classes 1 through 4. The basic difference between these two categories of switches is that a class 5 local switch (also referred to as an end office) provides the ability to directly interconnect or interface to a subscriber's terminal equipment while a tandem-type switch will only provide interconnection to other switching equipment or systems. Thus the local class 5 switch has the ability to terminate a call to one of its subscriber units while a tandem switch can only route calls to other destined nodes and can never act as a final call delivery point (see Fig. 3.3). The routing of a call through the network occurs according to a "three-level limit rule." This call routing process is described in Table 3.2 and Fig. 3.4.

In some cases the call will get routed to the highest class 1 office switch, where it is routed to another class 1 office switch. The latter class 1 switch will be in a direct vertical line, as far as switch interconnection is concerned, with the terminating class 5 office switch. The routing of the call between the initial class 1 office switch and this final class 1 office switch may take place by passing through a number of intermediate class 1 office switches.

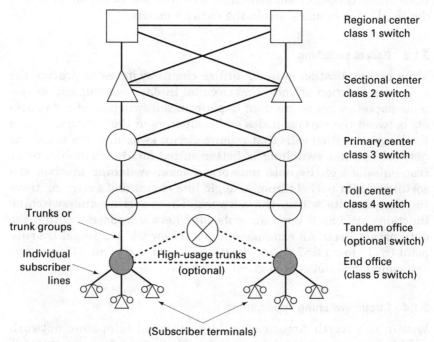

Figure 3.3 North American telephone switching hierarchy.

TABLE 3.2   Call Routing through the North American Telephone Switching Hierarchy Using the Three-Level Limit Rule

Step A	When a subscriber connected to a class 5 office (the call originator) attempts to call another subscriber connected to a different class 5 office (the destined call termination), a route is sought over a set of existing facilities (see the definition of facilities below) directly interconnected to the terminating class 5 office switch.
Step B	If no direct interconnecting facilities exist, or they are all currently being used (in a busy state), then an attempt is made to route the call through the tandem office switch.
Step C	If no tandem office exists or no route is available to the desired class 5 office through this node, then the call will be routed up one level of the PSTN hierarchy to the toll center switch.
Step D	Upon receiving the call from the originating class 5 office switch the toll center switch determines whether it can deliver the call to the terminating class 5 office switch. If it cannot perform this routing, it attempts to send the call to the toll center switch which serves the terminating class 5 office switch.
Step E	If there are no routes to this toll center node, an attempt is made to route the call to the class 3 office switch which serves that toll center.
Step F	If this routing step also fails, the call is passed up to the class 3 office switch directly above this switch in the hierarchy. At this point the process begins again for routing the call to its final destination.

The routing of a call through a network of switches, based upon a hierarchy design, is completed with the intent to establish a call connection to a switch that is in direct vertical line with the terminating switch. The "three-limit rule" is derived from the restriction that each transit switch has only three levels from which to choose to route a call. It first tries to deliver the call to a level below itself. If this step is not possible, the switch attempts to route the call to another switch at its own hierarchy level. Should this step fail, the call is routed to a switch one level higher than itself. These three levels are all the possible routes available to the transit node. No network switch can directly pass a call two levels up or down in the hierarchy. See Fig. 3.4 for this routing example. Within the telecommunications industry the class 5 switch is sometimes referred to as a local switch local exchange, or an end office, and the tandem switch is sometimes referred to as a toll exchange.

Within the PSTN traffic is switched through individual nodes using the interconnecting trunk groups. The offices and trunk groups are organized in a hierarchical manner. Call routing within this network begins at the subscriber unit connected to the class 5 end office with the customer initiating a call to another subscriber. If the call is to be placed to another subscriber that happens to be connected to the same end office, the call is deemed a local call and no additional

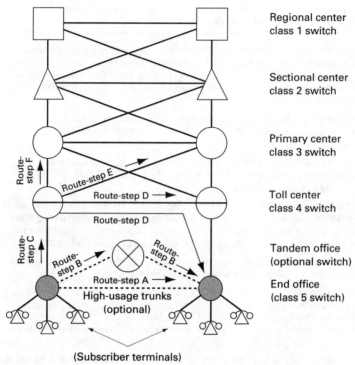

Figure 3.4 Call routing example in the North American telephone switching hierarchy.

charges are incurred by the originating party. If, however, the call is placed to a subscriber that is connected to a different end office, it must be routed to a toll exchange according to the routing example described above, and the call will typically incur additional charges. These descriptions apply in general. Exceptions may well exist; however, they will serve as an introduction to this basic material. As a side note, a cellular mobile switch is considered to be equivalent to a class 5 office switch in the North American hierarchy.

3.2 Components and Concepts of the Voice Network

Many types of interconnection and methods of transmission are available for providing communication between the various nodes in a wireless network. The following section discusses some of these designs and techniques for readers to gain a basic understanding of these topics to prepare themselves for working within the wireless industry.

3.2.1   Pulse-code modulation (PCM)

PCM is a process for digitizing and encoding analog signals and is used, along with other processes, in T-carrier-based transmission systems. These T-carrier-type systems are the most widely used for interconnecting switches within PSTN and wireless networks. Figure 3.5 shows a functional block diagram of the PCM process.

The PCM process step 1—sampling. The first step in the PCM process is called sampling. It involves choosing and measuring points along the analog speech signal. The actual measured values are called samples. They produce a series of pulses which represent the amplitudes of the various signals at specified known times. In order for this process to take place, the rate at which to sample (time intervals between successive measurements) the original analog voice signal needs to be determined. This rate is dependent upon the Nyquist theory and upon the voice-grade bandwidth as determined by the telecommunications industry. The Nyquist theory is stated: If the sampling rate is great enough (two times the highest frequency of the signal to be sampled), then it is possible for the receiving equipment of a communications system to reconstruct the original analog signal based upon the sampled signal data. The industry bandwidth limits for voice-grade analog signals range from 300 to 3000 Hz. The actual upper limit, however, is set at 4000 Hz. Given the above rule and

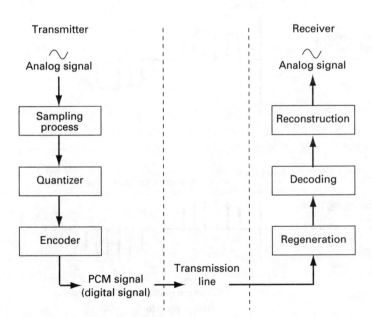

Figure 3.5   PCM functional diagram.

data, the sampling rate is therefore twice the bandwidth of the upper limit: 4000 Hz times 2 equals 8000 samples per second.

Once the sampling process is conducted upon the analog signal, it yields a digital signal that represents only one-half the original signal. This digital signal will not resemble the voice frequency signal. Instead it will appear as a series of pulses whose height will be the same as the original analog signal at the points where the samples were measured (see Fig. 3.6). The analog signal which has been divided into a series of short sampled pulses is referred to as having been pulse-amplitude modulated (PAM).

The PCM process step 2—quantizing.   The next step is the quantization process. This process takes the amplitudes of the samples generated in step 1 and quantizes them so that they can be represented by an eight-digit binary digital code. The quantization step, or quantizer, assigns the amplitude ranges of the PAM signal into a finite number of amplitude values. This is completed by dividing the total amplitude range into intervals. Numerical values are assigned to each interval, depending upon its distance (positive or negative) from the time axis. These values form quantizing intervals into which the PAM samples can fall. Each sample is assigned the value of a quantizing interval depending upon which one it is closest to in height. Since the heights

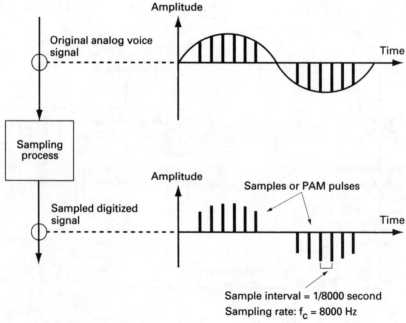

Figure 3.6   PCM sampling process.

of the PAM samples rarely match the heights of the quantizing intervals exactly, there will be a certain amount of error or rounding of the samples. This error (called quantizing error) is irretrievable and adds noise to the signal (called quantizing noise). This noise is heard as a hissing in an actual telephone (see Fig. 3.7 for a graphical representation of this process).

The μ-255 or μ-Law is the Bell standard quantization process for use in the United States. This specification allows for better resolution of the lower quantization levels, which are more predominant in PAM voice signals. The European standard is the A-Law, which provides a slightly better signal-to-noise ratio for small signals but has a higher idle-channel noise level.

The PCM process step 3—encoding.   The final step of the PCM process is called the encoding step. This step takes the output values of the quantizer and converts them into a binary serial data stream that is capable of being transmitted over a single transmission line (see Fig. 3.8). This process is then reversed at the receiving end of the communication system to yield the original analog signal.

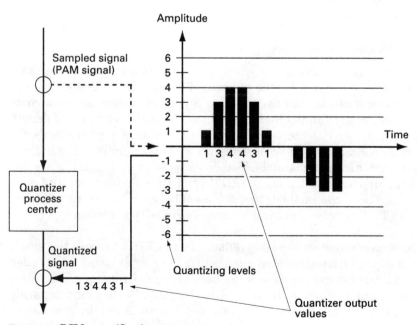

Figure 3.7 PCM quantification process.

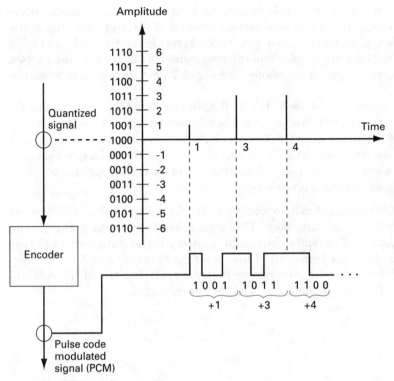

Figure 3.8 PCM encoding process.

3.2.2 Multiplexing methods

Multiplexing is a process by which a number of communication channels can be combined for the purpose of being transmitted over a single communication channel that has a greater transmission capacity (broader band) than the individual input channels. A process of demultiplexing is used at the opposite end of the multiplexed transmission facility to separate and reconstruct the original signals of the individual communication channels. Two basic types of multiplexing are frequency-division multiplexing (FDM) and time-division multiplexing (TDM). These two methods are discussed below, with emphasis placed on the TDM process since it is used in T-carrier-type systems.

Frequency-division multiplexing (FDM). In the FDM process the single large-capacity transmission channel (facility or system), also called broadband channel because of the broad range of signals it can transmit without distortion, is divided by frequency into many separate individual channels. These constituent channels are in continuous operation; i.e., signal transmission is never stopped. Signals from these channels are all transmitted at the same time; however, each uses a different carrier frequency. Given a standard voice analog signal that ranges from 0 to 4 Hz, the FDM process would take this individual analog signal and superimpose it onto a higher-frequency carrier signal by a process called amplitude modulation. At this point the individual analog signal is assigned to a specific frequency slot or one of the separate channels mentioned above that is associated with the high-capacity transmission channel or system (broadband channel or system). All individual frequency slots or channels are transmitted over the same physical piece of equipment (facility) at the exact same time. At the opposite end of the transmission system the process is reversed (FDM demultiplexing) and all the original voice analog signals are restored.

Time-division multiplexing (TDM). In the TDM process the large-capacity transmission channel is divided up into time intervals called time slots instead of the frequency slots mentioned in the FDM process. Next PAM samples from the PCM process described in Sec. 3.2.1 are interjected into a single high-capacity (and higher-speed) transmission channel (see Fig. 3.9 for a graphical representation of this process). This process of interjecting PAM samples (pulses) of each individual input channel into the high-speed transmission channel occurs once every sample cycle. The total PAM samples or pulse signals embedded into the high-speed channel during one cycle is called a frame. On the receiving end of the transmission facility this process is reversed (demultiplexed) and the original signals are reconstructed.

The final output of the TDM process is a TDM-PAM signal. Taking this concept further, if the TDM process were conducted not using the PAM samples but the PCM encoded word representing the PAM sample, then the resulting final output would be a TDM-PCM signal. This subprocess is called time slot interleaving. Most TDM systems used in the telecommunications industry are of the TDM-PAM system type (see Fig. 3.9).

3.2.3 T-carrier systems

T-carrier systems provide digital two-way transmission of voice, data, or video signals over a single high-speed circuit. These systems have been used by telephone companies and other carriers for interconnecting network switches since the 1960s. They have evolved considerably since then, but the basic concepts still apply. The following is a brief description of these systems to "whet your appetite."

The actual data transmission rate of a single T-carrier-based system (referred to as a T-1 carrier) is based upon the bandwidth (range

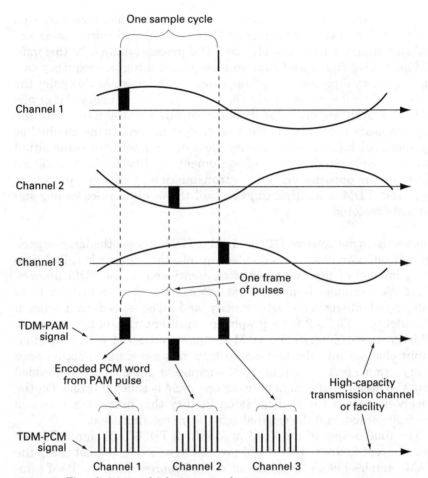

Figure 3.9 Time-division multiplexing example.

of frequencies) utilized to transmit one digitized voice signal. This signal is referred to as DS-0 (digital signal, level zero). The bandwidth of this DS-0 signal is equal to 64,000 bits per second. The initial T-1-based systems transmitted 24 voice channels over two pairs of twisted cables at a rate of 1,536,000 bits/s. Later systems introduced a necessary control bit for synchronizing T-1 multiplexing equipment. This additional bandwidth brought the T-1 carrier transmission rate up to 1,544,000 bits/s (also denoted as 1.544 Mbits/s). This is now known as DS-1, the first signal level of a T-1 carrier system. Table 3.3 presents the various signal levels and transmission rates for T-carrier-based systems.

The original multiplexing equipment of the T-carrier-based systems were referred to as channel banks or "D-banks." The first versions

TABLE 3.3   T-1 Carrier Systems Data

Signal level	Carrier system	Number of T-1 systems	Number of voice circuits	Mbits/s
DS-1	T-1	1	24	1.544
DS-1C	T-1C	2	48	3.152
DS-2	T-2	4	96	6.312
DS-3	T-3	28	672	44.736
DS-4	T-4	168	4032	274.760

produced were the D1 line of channel banks. The early framing structure was known as the D1 framing format. As these systems evolved, more improved framing formats were developed. From the original D1 format emerged the D1D, D2, D3, D4, and the current extended superframe format (ESF). With these newer systems came an improvement in overall performance, error correction, and diagnostic techniques.

T-carrier systems can be combined to form larger systems with greater traffic-handling capacity and transmission throughput (speed at which data can be transmitted over the given system). An example of this combining of T-carrier systems is shown in Fig. 3.10. T-carrier-type systems are used in wireless networks to interconnect cell sites to the mobile switches and to interconnect mobile switches to other mobile switches for the purpose of providing voice and data communications. In Fig. 3.11 a simple cellular wireless network is presented. In this network cell site A is connected to switch 1 in MTSO A by use of a single T-1 span from the cell site to the PSTN. Then, by prior arrangement between the PSTN and the wireless carrier, this T-1 span is multiplexed into the DS-3 facilities into MTSO A. Once inside MTSO A, the DS-3 level is demultiplexed down to a T-1 level signal for direct interconnection and input into switch 1. Similarly, cell site B is interconnected to switch 2. In this example the 24 channels of the T-1 span interconnecting cell site A to switch 1 are broken down into 22 voice channels and two data channels. The two data channels are used for data communications between the switch and cell site for control of the mobile calls and other mobile functions at that site. Also in Fig. 3.11, the DS-3 between MTSO A and the PSTN and between MTSO B and the PSTN provides a means of interconnection between these two switches.

3.2.4   PSTN interconnection types

Various types of facilities are used throughout the PSTN based upon the particular application within which they are used. Table 3.4 lists these facilities and some possible applications. The definitions of the

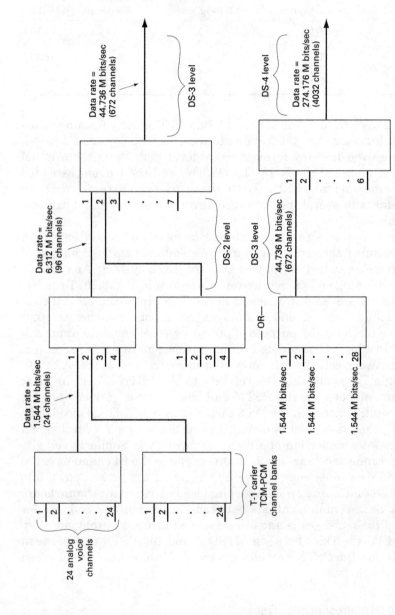

Figure 3.10 Example of combining of T-carrier-based systems. (Note: 1.544 Mbits/s means 1,544,000 bits transmitted in one second. Key: DS = data signal, TDM = time division multiplexing, PCM = pulse code modulation.)

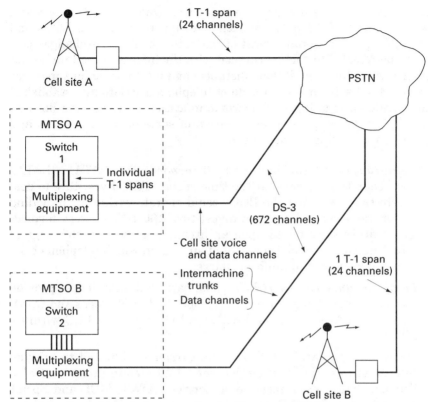

Figure 3.11 Example of wireless network using T-carrier systems.

TABLE 3.4   PSTN Facilities and Applications

Facility type	User signaling	IXC-LEC circuit type	Application
A	DTMF	Line circuit	Switch to switch
B	DTMF	Trunk	950 type calls
C	DP/DTMF	Trunk	AT&T interconnection only
D	DP/DTMF	Trunk	Equal access

DTMF and DP signaling, as well as other signaling types, are given in Sec. 3.2.6.

3.2.5 Telephone carrier types and entities

When the Modified Final Judgement (MFJ) of 1982 was passed, it required AT&T to divest itself from the Bell Operating Companies (BOCs) under its control at that time. The nationwide phone network that was designed, constructed, and operated as a single system was now expected to be divided into two parts. One part of this breakup would consist of the network exchanges (switches) of the BOCs and the other portion would consist of the network of interexchanges provided by AT&T. This action coupled with the intrastate and interstate restrictions plus the toll (long distance) and local restrictions together formed the basis for a multitude of telephone carriers all established to provide various types of telecommunication services. The list and description below give an explanation of some of these carriers and their corresponding service offerings.

Definition of a local access and transport area (LATA).   LATAs were boundaries formed at the time of the AT&T divestiture to designate the areas that the BOCs could operate within for providing telephone service. From the divestiture 160 LATAs were formed. These areas were based upon an exchange area from the original AT&T network, which represented an area in which telephone calls received a single and uniform charge.

Local exchange carrier (LEC).   The carrier designated to provide telecommunication service within a specified local access and transport area (LATA) or areas (LATAs). A BOC then is a LEC within a LATA.

Interexchange carrier (IXC).   The carrier designated to provide telecommunication service between LATA boundaries and also between countries (international service). AT&T, MCI, and Sprint are examples of an IXC.

Local call.   A local call is defined as the type of telecommunication service that is offered by the LEC for delivering calls within a LATA. This type of call does not require the additional charges associated with a toll-type call.

Toll call.   A toll call is defined as the type of telecommunication service that is offered by the IXC for delivering calls outside a LATA and outside the country. Additional charges are applied to this type of call owing to the larger number of network switches and facilities required for its completion.

As an example of these concepts refer to Fig. 3.12 and the following call-routing descriptions for the various call types.

1. If subscriber X calls subscriber Y, the call is routed by use of LEC N's switch to subscriber Y, and the call is deemed a local call.

2. If subscriber X calls subscriber S, the call will get routed to LEC N's switch where it will then be routed to IXC Z's switch 1. From

there the call will be routed to IXC *Z*'s switch 2 by use of the inter-LATA trunks. Once in LATA B the call is routed to LEC Q switch and then on to subscriber *S*. This call is deemed an inter-LATA toll call.

Note in Fig. 3.12 that a number of LECs and IXCs can be involved in a call depending on the network design, subscriber's choice in long-distance carrier (IXC), etc.

3.2.6 Common telephony terms and equipment

Channel bank. Any of a variety of terminal equipment used as a transmission system for the purpose of multiplexing individual channels using FDM or TDM techniques.

DAC (digital access cross-connect). This equipment provides the ability to automatically interconnect (without manual wiring by use of software-controlled hardware) individual channels of a T-1 span on a one-by-one basis. This functionality allows for grooming and filling of the input T-1 spans to the DAC unit. The equipment can also be used with T-3 type facilities.

Data service units (DSUs). This equipment amplifies an incoming signal from the PSTN for retransmittal to the application equipment. In many cases this application equipment is a system switch.

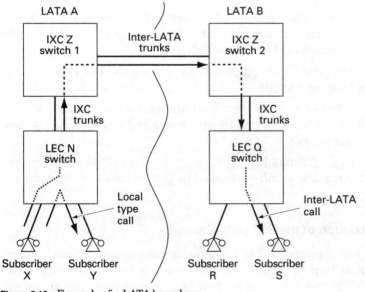

Figure 3.12 Example of a LATA boundary.

Dial-pulse signaling.   A method of signaling where direct current flow through the loop of a calling telephone is alternately interrupted, then restored. For example, the rotary dialing of the digit 7 generates seven dial pulses (interruptions) in current flowing through the loop of the calling telephone.

DTMF (dual-tone multifrequency).   A method of line address signaling using two of seven in-band tones.

E & M signaling.   A method of conveying supervisory and address signaling on a trunk by means of two signaling leads called the "E lead" and the "M lead." Signals are transmitted on the M lead and received on the E lead. There are several varieties of E & M signaling which differ depending upon the application.

Erlang.   The unit of measure for telephone traffic used throughout the telecommunications industry. A single circuit which is occupied continuously over a 1-hour period is equal to one erlang.

Facility.   A facility is a medium of transmission which connects two fixed points in a network for the purpose of providing a means of communication between network elements. This medium could be a pair of copper wireline circuits, a fiber-optics circuit, a microwave radio circuit, etc.

Glare condition.   Glare is a condition that exists on a two-way trunk when it is seized (chosen by a switch for transmitting information) from both ends.

MF signaling.   The high-speed (in-band) 2 of 6 analog tone trunk signaling system that is used primarily for address signaling in the North American intertoll network.

POTS.   This stands for plain old telephone service, the traditional wireline telephone service.

Trunk.   A trunk is a facility which connects a subscriber unit to a network switch or a switch-to-switch connection, a switch-to-tandem connection, etc.

Trunk group.   Trunks having similar characteristics and interconnecting points are usually arranged in groups called trunk groups.

3.3   Description of the SS7 Data Network

Signaling system number 7 is a popular data communications transfer protocol used to provide out-of-band signaling for processing calls in both the PSTN and cellular networks. It is an improved method of signaling over the traditional in-band signaling methods of MF, DTMF, and DP. This protocol has three major advantages over in-band signaling. The first advantage is the improved postdial delay for faster call setup times. This network improvement is very noticeable to the end user. The second advantage is the ability to signal in the reverse direction, improving communication between various nodes in the network. A third advantage of SS7 over conventional in-band signaling is the improvement in network fraud prevention because the data (voice) messages and signaling messages are sent over separate routes.

3.3.1 SS7 network components

Signaling service points (SSPs) and signaling points (SPs). SSPs and SPs serve as the connecting points for end subscriber units such as landline and mobile phones. These nodes perform call processing on calls that originate, terminate, or tandem at the location.

Signal transfer points (STPs). STPs transfer SS7 messages between interconnected nodes based upon information contained in the SS7 address fields.

Signaling control points (SCPs). SCPs operate in a manner similar to the SSPs and SPs but process only database information functions. A *home location register (HLR)* is an example of an SCP.

Access links (A links). Access links carry SS7 messages between SSPs and STPs and between STPs and SCPs.

Bridge links (B links). Bridge links carry SS7 messages between STPs in different regions of the same network.

Control links (C links). Control links carry SS7 messages between mated STPs.

Diagonal or quad links (D links). Diagonal or quad links carry SS7 messages between STP in the network pairs.

Extended links (E links). Extended links carry SS7 messages between SSPs and remote STPs.

F links. F links carry SS7 messages between SSPs.

Linkset. Two or more links connect adjacent nodes and share the same routing information.

See Fig. 3.13 for more details about the use of these components in an SS7 network.

3.3.2 SS7 data network protocols

A number of protocols can be supported by the SS7 network. While SS7 network protocol provides the reliable transport of data messages

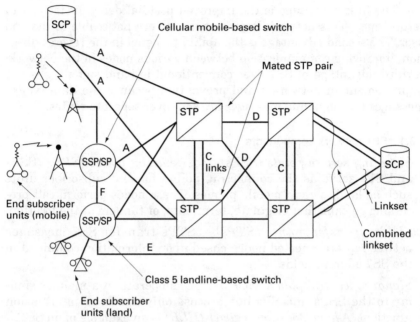

Figure 3.13 Example of a basic SS7 network.

between network nodes, additional protocols are used to build and decode these various messages based upon the particular application in the network. The use and function of these additional protocols of an SS7 network follow the open-system interconnection model shown in Fig. 3.14. This model consists of seven layers as described below.

1. *Physical layer.* This layer provides the electrical and mechanical interface to the network transmission facilities. It defines the protocol for actually transmitting a stream of serial data bits between two network nodes.

2. *Data link layer.* This layer provides control for the transmission, framing, and error correction functions over a single transmission data link facility.

3. *Network layer.* This layer provides the functions to establish clear, logical, and physical connections (if required) across the network. Included in these functions are network routing (addressing) and flow-control functions across the computer to network-type interfaces.

4. *Transport layer.* This layer provides independent and reliable message interexchange functions to the upper three application-

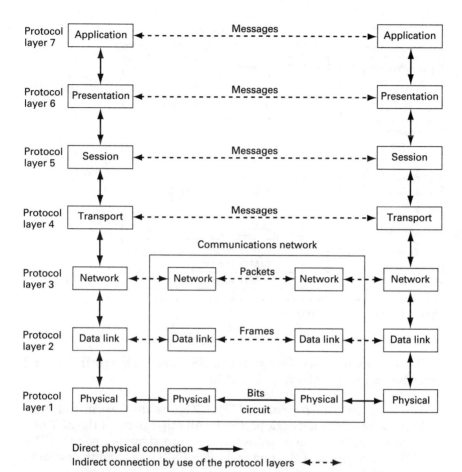

Figure 3.14   Open-system interconnection model.

orientated layers. It acts as an interface between the bottom three layers and the upper three layers.

5. *Session layer.*   This layer provides the functions that control the data-exchange process. It sets up and clears communication channels between two communicating network entities.

6. *Presentation layer.*   This layer provides the syntax translation between two application communications over the network.

7. *Application layer.*   This layer provides the user interface to a variety of network services and also manages the communication between applications on two communicating systems. Example applications include file transfer access and management (FTAM) and electronic mail.

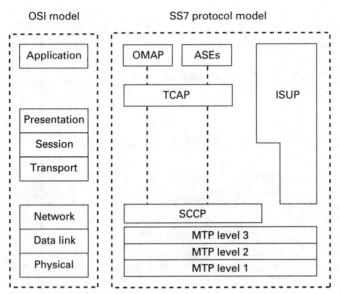

Figure 3.15 SS7 protocol structure.

The six layers pertaining to the SS7 network are listed and described below and shown in Fig. 3.15.

1. *The message transfer part (MTP).*   This is the lowest and most basic layer in the network protocols. All other parts of the SS7 network use MTP for basic transport between signaling points (SPs). MTP is comprised of the physical (data link) layer and the network layers of the OSI model.

2. *The signaling connection and control part (SCCP).*   This corresponds to the transport layer of the OSI model. It provides support for both connectionless and connection-oriented network services.

3. *The telephone user part (TUP).*   This part provides for call setup and teardown under the SS7 network protocol.

4. *The ISDN user part (ISUP).*   This part provides interoffice ISDN services and uses either SCCP or MTP for message transport.

5. *The transactions capabilities part (TCAP).*   This part uses SCCP for message transport. TCAP supports the transfer of messages not associated with circuit control.

6. *The operations maintenance and administration part (OMAP).* This part provides the application protocols to monitor and coordinate all the network resources.

The interswitch standard utilized for providing automatic call delivery and handoffs in the mobile network is the EIA/TIA IS41 protocol. This protocol resides in the TCAP portion of SS7 network structure and uses the SCCP layer as a message transport. This protocol standard is still in development but is currently used in North American Markets.

3.4 Auxiliary Cellular Systems

In addition to the base systems that provide wireless communications a number of other support systems are used by the operators to assist in the management of the network and to offer special services to their customers. A brief description of these types of systems is given below for review. Be aware that many more such systems are available and in use in the industry today. It is suggested that you compile a list of all the systems currently in use in your system and obtain as much detail about each product as is available. This information will become a valuable manual for company personnel to reference. Be sure to include not only the engineering and operations departments but the marketing, information services, and customer care departments as well.

3.4.1 Validation systems

Validation systems are required by the cellular system to provide a means of checking the validity of the mobile subscriber. A valid subscriber (a paying customer in good credit) is allowed to use the system. A mobile subscriber who is not a valid user will be denied service on the system.

There are two basic types of validation systems, postcall and precall. The initial cellular systems used postcall validation systems for determining if the mobile subscriber was authorized to use the system or not. Once a mobile user made a call on the system the system would begin to analyze the data and determine their status in the system. These systems worked well in the beginning, but soon a more advanced system was needed. The next generation of validation systems took advantage of the SS7 networking protocol and registration process of the mobile system to validate a user prior to an initial call. This system has helped reduce the number of unauthorized mobiles using the system, which in turn reduced the total amount of system fraud and allowed for the network facilities to be available for legitimate paying subscribers.

3.4.2 Voice mail system

Most cellular systems offer answering-machine-type services to their mobile subscribers. This is a very popular and useful service to the mobile customer, but like any system it must be engineered, maintained, and expanded when the use of the system begins to reach a critical upper limit. This system should be treated much like the network switches in that it has a capacity limit on the number of subscriber accounts that can be stored in the individual nodes and in the system as a whole. A voice mail system will also have a limit for the number of calls that it can process at a given time. Therefore, system usage statistics should be collected during the system busy hour and analyzed much like the usage on the network switches.

3.4.3 Value-added services

Value-added services are additional features offered by the cellular carrier to their mobile subscribers for enhancing their service. Examples of these services are:

- Information line for traffic, sports, trivia, etc.
- Data services for using cellular modems in the system
- 911 service
- 611 customer service

The concepts presented in this chapter are very basic and essential to any network manager or engineer working on a cellular system. Further reading in the sources listed below is recommended to gain a more detailed, broader knowledge base of telephony for use in designing and operating live cellular systems.

References

1. AT&T Bell Laboratories, *Engineering and Operations in the Bell System*, 1984.
2. Bellcore, Bell Communications Research Specification of Signaling System, vol. 1, no. 7, TR-NWT-000246, Issue 2, June 1991.
3. Ericsson Radio, Inc., *MTSO Operations and Maintenance Course Manual*, vol. I, Richardson, Tex.
4. Hebuterne, Gerard, *Traffic Flow in Switching Systems*, Artech House, Inc., Norwood, Mass., 1987.
5. INET, Inc., *SS7 Overview*, Richardson, Tex.
6. Lee, William C. Y., *Mobile Cellular Telecommunications Systems*, McGraw-Hill, New York, 1989.

RF Design Guidelines

No true RF engineering can take place without some form of RF design guidelines, whether formal or informal. However, with the level of complexity in wireless communication systems rising every day the lack of a clear definitive set of design guidelines is fraught with potential disaster. While this concept seems straightforward and simple, many wireless engineering departments when pushed have a difficult time defining what exactly their design guidelines are.

This chapter covers design guideline philosophies to utilize in building, expanding, and operating a wireless communication system. What should be presented at a design review, signatures required and when, plus design change orders are covered. These guidelines can be easily crafted into specifics for your company.

The actual format and how it is conducted should be structured to facilitate ease, documentation, and minimization for formal meetings. For most design reviews a formal overhead presentation is not required. Instead a sit-down with the manager of the department is in most cases the level of review needed. The important point is that another qualified member of the engineering staff should review the material to prevent common or simple mistakes. Ensuring that a design review process is in place does not eliminate the chance of mistakes. Design reviews ensure that when mistakes do take place the how, why, and when issues needed to expedite the restoration process are already in place.

It is highly recommended that RF design guidelines are reasonably documented and updated on a yearly basis, at a minimum. The use of design guidelines will facilitate the design review process and establish a clear set of directions for the engineering department to follow. The guidelines will also ensure a consistent approach to designing and operating the capital infrastructure that has or will be put into place within the network.

4.1   Cell Site Design

While this is not necessarily the first step in any design process, it is one of the most important for the RF engineering department. The cell site design is critical for the RF engineering department because it is where the bulk of the capital is spent. The guidelines listed below can be utilized directly or modified to meet your own particular requirements.

The use of a defined set of criteria will help facilitate the cell site build program by improving interdepartmental coordination and providing the proper documentation for any new engineer to review and understand the entire process with ease. Often when a new engineer comes onto a project all the previous work done by the last engineer is reinvented primarily owing to a lack of documentation and/or design guidelines.

The cell site design process takes on many facets, and each company's internal processes are different. However, no matter what internal process you have, the following items are needed as a minimum.

1. Search area

2. Site qualification test (SQT)

3. Site acceptance

4. Site rejection

5. FCC guidelines

6. FAA guidelines

7. Planning and zoning board

8. EMF compliance

4.1.1   Search area

The definition of a search area and the information content provided is a critical first step in the cell site design process. The search area request is a key source document that is used by the real estate acquisition department of the company. The selection and form of the material presented should not be taken lightly, since more times than not the RF engineers rely heavily upon the real estate group to find a suitable location for the communication facility to exist. If the search area definition is not done properly in the initial phase, it should not be a surprise when the selection of candidate properties is poor.

The search area should follow the design objectives for a given area following the growth study format layout in Chap. 9. It should be put together by the RF engineer responsible for the site design. The final paper needs to be reviewed and signed by the appropriate reviewing process, usually the department manager, to ensure that there is a check and balance in the process. The specifications for the search area document need to meet not only the RF engineering department requirements but also the real estate and construction group needs. Therefore, the proposed form in Fig. 4.1 needs to be approved by the various groups, but issued by the RF engineering department. It is imperative that the search area request undergo a design review prior to its issuance. The proposed format that should be followed for the search area request is shown in Fig. 4.1.

Referring to Fig. 4.1, the following comments need to be made. The map included with the search area request needs to include as much information as practical for the real estate acquisition group to help locate the proper sites. The map used in this form will minimize the number of dud sites that are presented to RF engineering for consideration in the system design. It should also include area-specific information.

The information needed varies from location to location. Its content is different if the proposed site is in a very rural area or in a dense urban environment. The variations you can use for the map format are largely dependent upon your design criteria for the site. If the search area ring is very defined, as is the case with mature systems, it is imperative that the adjacent existing sites and search areas are identified on the map itself. The rationale behind including adjacent sites and the current search areas will better define options available to the real estate acquisition group.

A propagation plot for the search area request form needs also to be generated and included with the search area folder maintained by RF engineering. The objective behind having a propagation plot generated is to help define the search area and coverage rings put forth in the map provided. The propagation plots will be used as part of the site acceptance procedure listed later. The propagation plots are one of the steps taken to ensure the proposed site meets its desired objective. On the search area request form the search area code should be identified along with its capital funding number. The capital funding number and the search area code can and should be the same.

The on-air target date is meant to identify when the site is needed to be placed into commercial service. The on-air date should match the dates put forth in the system growth plan defined in Chap. 9. The

Company "X"
RF Engineering Department
Search Area Request

Map

Search area code: _____   Capital funding no.: _____

On-air target date: _____/____/_____

Search area type: Capacity, Coverage, Frequency plan, Competitive, New technology

Design objective: (description of sites key objectives)

Cell configuration: Omni, 3 sector, 6 sector, other

Type of infrastructure: (equipment type)

Total antenna count: (number and type)

AGL of antennas: _____ ft/meter

ASML of antennas: _____ ft/meter

Comments: Revelant comments about the search area

Search Area Request			
Document No.:		Date:	
Comments	Design	Reviewed	Rev.

Figure 4.1   Search area request.

on-air date's purpose is to help prioritize the internal resources of the company by helping define the importance of the site.

The search area type is meant to notify internal and external departments what the actual purpose of this site is. There are functionally five main groupings for cell site types: capacity, coverage, frequency planning, competitive, and new technology. The type of search area this site represents is important for everyone involved to understand.

The design objective for the site is a critical issue with the search area request form. Stating what the design objective is for the site will assist in the site acceptance design reviews. Stating the design objective will also assist other engineers in determining the rational behind why the site is needed. The design objective should match the goals set forth in the system growth plan put forth in Chap. 9.

The cell site configuration and infrastructure type is meant to assist the real estate acquisition department in determining the parameters for the physical site. The physical parameters for a potential site are important to have defined in advance since this directly impacts the equipment space and power requirements. The floor space requirements obviously are different if the site is a microcell or a macrocell. The total number of antennas and their type are also important to define in advance for the site search. If a site requires three antennas versus sixteen antennas the discussion in the approach the real estate acquisition group takes with the potential landlord is different.

Listing the above ground level (AGL) value and above mean sea level (AMSL) will help define the site location options for the real estate acquisition group. The AGL will help structure the search for suitable properties that fit the design parameters. Sometimes sites are available that are 10 m tall and have a willing landlord and permit accessibility. However, the design specification calls for a 30-m antenna height, disqualifying the site for consideration and resource expenditure. If a site needs to be 30 m tall, about 100 ft, its AMSL is important to define since it might be possible to have the antennas at the 30-m height but the location of the property is in a gully, requiring a 50-m tall antenna installation.

The comment section of the form in Fig. 4.1 is meant to provide an area for the designer to specify any particulars desired for the site. On the search area request form itself is a section for documentation control, which is meant to track who issued the search area request, what revision it is, and the dates associated with it. Also on the document control portion of the form is a section for the design reviewer to sign off on the request. It cannot be overstressed that the information on the search area request form will largely define the success or failure of the property search. All the search area requests need to undergo a design review.

4.1.2 Site qualification test (SQT)

The site qualification test (SQT) is an integral part of any RF system design. Every site needs to have some form of transmitter or site qualification test conducted at it. The fundamental reason behind requiring a test to assure the site is a viable candidate before a large amount of company capital is spent on building the site. This test is also required to make sure the site will operate well within the network. The financial implications associated with accepting or rejecting a cell site necessitate a few thousand dollars expended in the front end of the build process. If a site is accepted that will not perform its intended mission statement additional capital will need to be outlaid to accomplish it. Several stages need to be done in this process. It is very important that the SQT be performed properly, since this will determine whether over $500,000 to $1 million will ultimately be spent on the facility.

It is strongly recommended that the RF engineer responsible for the final site design visit the location prior to any SQT. This site visit will facilitate several items. First the engineer will now have a better idea of the potential usefulness of the site and its ability to be built and can provide more accurate instructions to the testing team. The RF engineer should not design the test on the fly by instructing the testing team on that day where to place the transmitter and which routes to drive. The desired approach is to have the engineer determine where to place the transmitter, either which part of the tower or rooftop, or the location for the crane. The RF engineer then puts together a test plan, identifying the location of the transmitter antenna, the ERP, drive routes, and any particular variations desired. The test plan is then submitted to the manager of the department for approval and is passed to the SQT team.

The rationale behind putting this step in the process is driven by the capital dollars needed to be spent and the system interoperability issues. The SQT is a very critical step in the process and needs to be well thought out in advance. Improperly defining the test routes or transmitter locations can lead to a poorly performing site being accepted or a potentially well-performing site from not getting accepted. The driving issue here as always is that a well-planned test will save time and money many times over.

The proposed form to use for the site qualification test (SQT) process is shown in Fig. 4.2. The format of the SQT form needs to directly match the input requirements of the RF engineers, the SQT

Date: ___/___/___

Site Qualification Test

Search area code: _____   Capital funding no.: _____

Address of test site: _____

_____

_____

_____

Site contact   Name: _____

Phone no.: (____)_____

Requested test date: ___/___/___

7.5 minute map: _____

Test antenna: _____

Test antenna height: _____

Test ERP: _____ watts

Test frequency/channel: _____

Test mounting information: roof, tower, crane, water tank, misc.

Rigger required: _____ (Y/N)

Test location sketch attached: _____

Test routes plan attached: _____

SQT team leader: _____

Postprocessing information:

Scale: _____

Color code: _____

Data reduction method: _____

SQT test calibration date: _____

SQT transmitter calibration date: _____

Site Qualification Test		
Document No.:	Date:	
Test No.	Design	Reviewed

Figure 4.2 Site qualification test.

measurement team, and the group responsible for postprocessing the data.

It is imperative that the testing plan be reviewed and signed off as part of the design review process. The actual test conducted, antenna placement, physical routes traversed, and postprocessing criteria directly determine the viability of the location. This viability of a site is determined by completing the criteria set forth. A poorly designed test plan can cause either the failure of a site acceptance even though it may meet the needs set forth in the growth plan on the acceptance of a poor cell site. The test plan can also set the groundwork to approve a site even though the introduction of the site could have a negative impact on the network once it goes commercial.

The SQT needs to include a sketch of the test location defining where to place the actual test transmitter antenna. For example, when using a crane to place the transmitter antenna it is important to define where the crane should be parked and the test antenna height actually used. If the test location is not properly defined in advance, an error could occur in the test transmitter placement significant enough to pass or fail the SQT requirements. Figure 4.3 has several example diagrams of test transmitter antenna locations.

Figure 4.4 is an example of a drive test route for an SQT. The RF engineer working on the SQT must make certain that the drive test route defined matches the design criteria for their specified site. The individual line items of the SQT form shown in Fig. 4.2 are self-explanatory. The 7.5-minute map portion is meant to help specify the actual grid location of the test, which can be used as part of the archiving process for SQTs.

The test antenna type, height, and effective radiated power (ERP) obviously should be defined in advance to ensure the testing is conducted in accordance with the design specifications. The calibration requests on the form are meant to ensure that the equipment used is within calibration. It is important to always check the calibration for any test equipment used to be certain it is within specification and thus will provide reliable measurements.

4.1.3 Site acceptance (SA)

Once a site has been tested for its potential use in the network it is determined to be either acceptable or not acceptable. For this section the assumption will be that the site is acceptable for use by the RF engineering department as a communication facility. It is imperative that the desires of the RF engineering department be properly communicated to all the departments within the company

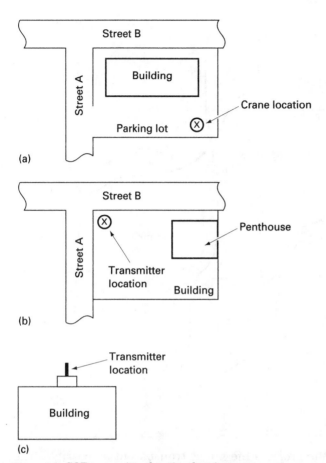

Figure 4.3 SQT transmitter location drawings.

in a timely fashion. The method of communication can be done verbally at first, based on time constraints, but a level of documentation must follow that will ensure that the design objectives are properly communicated.

The form listed below is meant as a general guide. It might need to be modified based on your particular requirements. Before the SA is released it is imperative that it go through the design review process to ensure that nothing is overlooked. The SA will be used to communicate RF engineering's intention for the site and will be a key source document used by real estate, construction, operations, and the various subgroups within engineering itself. It will also need to be sourced with a document control number to ensure that changes in

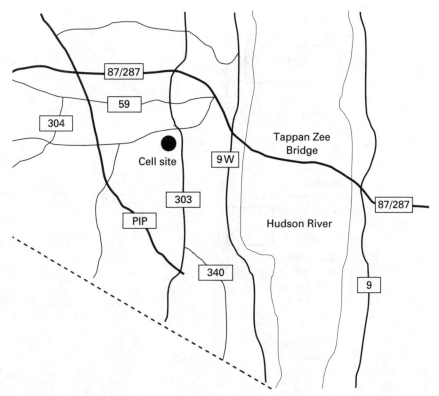

Figure 4.4 Drive route.

personnel during the project's life are as transparent as possible. The proposed site acceptance form is included in Fig. 4.5. It can be easily expanded to ensure that all the relevant information required within the organization is provided. Whatever the form or method ultimately utilized it is important to include the information listed in Fig. 4.5 as minimum requirements.

Most of the information included in Fig. 4.5 is self-explanatory. It is imperative that like all the other steps in the design process a design review and sign-off takes place establishing a formal paper flow. The SA needs to include a copy of the predicted propagation used to generate the search area request, a copy of the actual SQT plot utilized to approve the site for RF engineering, and a copy of the proposed antenna installation configuration. The proposed antenna configuration is used by equipment engineering and construction to evaluate the feasibility of the proposed installation. The antenna configuration drawing is also used as part of the lease exhibit information. A copy of

Site Acceptance Form

RF Enginnering

Search area code: _____   Capital funding no.: _____
Document no.: _____
Site address: _____

_____

Latitude: _____   AGL: _____
Longitude: _____   ASML: _____

FAA check completed: _____ (Y/N)

FAA lighting/marking required: _____ (Y/N)
FCC contour extension required: _____ (Y/N)

Existing transmitters on structure: _____ (Y/N)
If yes state freq., ERP, call sign: _____

Antenna structure: Roof, tower, monopole, water tank

Type of equipment room: Prefab, interior, exterior
Equipment type: _____
Equipment location: _____

Approximate coaxial cable run: _____

Type and quantity of antenna:

Sector	Type	Quantity	Orientation	ERP

SQT document no.: _____
Antenna structure sketch: _____
FAA check documentation included: _____
Propagation prediction plot included: _____
SQT plot included: _____

Comments: _____

_____

_____

_____

Site Acceptance Form		
Document No.:	Date:	
Test No.	Reviewed	Rev.

Figure 4.5 Site acceptance form.

the SA and the supporting documents needs to be stored in a secure central filing location so that all the information pertaining to this location and search is stored in one place and not distributed among many people's work areas.

4.1.4   Site rejection (SR)

In the unfortunate event that a potential site has been tested and is determined not to be suitable for potential use in the network a site rejection form needs to be filled out. The issuance of a site rejection form may seem trivial until there is a change of personnel and the site is tested again at a later date. The form serves several purposes. The first is that it formally lets the real estate acquisition team know that the site is not acceptable for engineering to use and they need to pursue an alternative location. The second is that this process identifies why the site did not qualify as a potential communication site. The third ties into future use where the SQT data are stored and can be used for future system designs when the site might be more favorable for the network.

It is recommended that the site rejection process include a design review with a sign-off by the manager. This is to ensure that the reasons for rejecting the site are truly valid and the issues are properly communicated. The form proposed in the SR (Fig. 4.6) needs to be distributed to the same parties that the SA would be sent to.

The fact that a site does not meet the design criteria specified at this time in the network design does not mean it will always be unsuitable. It is therefore imperative that the SQT information col-

---

Search area code: Date: _____

The (name of test location) was tested on (date of test) and did not meet the design criteria for the search area defined.

The test location did not meet the design criteria for the following reasons (state reasons):

RF engineer: _____

Engineering manager: _____

Figure 4.6   Site rejection form, RF engineering.

lected for this site be stored in the search area master file. This will assist efforts in later design issues that could involve capacity or relocation of existing sites to reduce lease costs.

4.1.5 Site activation

The activation of a cell site into the network is exciting. It is at this point that the determination is made for how effective the design of the cell site is in resolving the problem area. After the site acceptance process previously listed, numerous steps need to take place. The degree of involvement with each of these steps is largely dependent upon company resources available and the interaction required to take place between the engineering and construction departments. As a minimum these two groups should perform site visits together involving the group responsible for the cell site architectural drawings and overall design of the site structure. Regardless of the interaction between the groups, when it comes to "show time" it is imperative to have a plan of action to implement. The specifics for implementing a site activation plan are included in Chap. 6.

4.1.6 Design guidelines

The actual design guidelines that should be utilized by the RF engineers need to be well documented and distributed. The design guidelines, however, do not need to consist of voluminous amounts of data. They should consist of a few pages of information that can be used as a quick reference sheet by engineering. The sheet should be based on the system design goals and objectives set forth in the network and RF growth plan.

The actual content of the design guideline can vary from operator to operator. However, it is essential that a list of design guidelines be put together and distributed. The publication and distribution of RF design guidelines will ensure there is a minimum level of RF design specifications in the network. The proposed RF design guideline is shown in Fig. 4.7.

4.2 FCC Guidelines

With the rewrite of part 22 of the Code of Federal Regulations Title 47, CFR 47, the paperwork required by the FCC is significantly reduced. However, the reduced paperwork the government requires does not eliminate the fundamental issue of ensuring that the system remains within FCC compliance. Most current operators no longer

System name:
Date:

	RSI	ERP	Cell area	Antenna type
Urban	−80 dBm	16 W	3.14 km/sq	12 dBd 90 H/14E
Suburban	−85 dBm	40 W	19.5 km/sq	12 dBd 90H/14E
Rural	−90 dBm	100 W	78.5 km/sq	10 dBd 110H/18E
Voice channel/I	17 dB (90th percentile)			
Frequency reuse	$N = 7$			
Maximum channels per sector	19			
Antenna system:				
Sector cell orientation	0,120,240			
Antenna height	100 ft or 30 m			
Antenna passband	824–894 Mhz			
Antenna feedline loss	2 dB			
Antenna system return loss	20–25 dB			
Diversity spacing	$d = h/11$ (d = receive antenna spacing, h = antenna AGL)			
Receive antennas per sector	2			
Transmit antennas per sector	1			
Roof height offset	$h = x/5$ (h = height of antenna from roof, x = distance from roof edge)			
Performance criteria:				
Lost call rate, %	<2			
Attempt failure, %	<1			
RF blocking, %	1><2			

Figure 4.7 RF design guidelines.

need to file a 489 form to the FCC when they activate or modify a cell site in the network provided it does not change the outer contour of the network itself.

The potential problem with not being required to file for everything you do is a potential compliance issue in the future. It is highly recommended that an internal filing process with sign-offs be put in place following the same process that was in place prior to part 22 rewrite. The objective with this method is to ensure that the operator remains in compliance and that in the event of an audit the proper paperwork for the system is in place for every site.

FCC filings should be stored not only in the legal department but also with the site-specific files for each site. The master control number for each of the sites should be the key source reference number used followed by a date for a definer. The completion of the FCC form needs to include a sign-off process to, again, ensure that the overall process is being adhered to. It is also suggested that the proper paperwork is completed to ensure that no site is activated or modified without completing the FCC process.

The legal compliance issue is not the only reason to adhere to producing an internal filing for every site added or modified. In many communication companies the source documents specifying the current configuration of a site tend to be scattered about the engineering department in an unorganized manner. A procedure will prevent this from occurring and enable new and existing engineering and operations personnel to determine, without visiting the site, its major design attributes. Following the process will ultimately expedite the design or troubleshooting of the site.

Figure 4.8 covers what items need to be documented and the level of detail required. FCC form 600 should be used as the reference document, with particulars filled out for a new site. However, the information in Fig. 4.8 is the minimum you should have available. Failure to remain in compliance can and will result in fines to the company posing both internal and external problems.

4.3 FAA Guidelines

FAA compliance is mandatory for all the sites within a system. Although you do not need to file every site with the FAA it is necessary to ensure that every site is within compliance. The requirements for ensuring compliance are listed in CFR 47 part 17. Verification of FAA compliance should be covered during the design review process. If a site does not conform within the FAA guidelines, a potential

Date:

Cell name:

Activation/change date:

Site revision number: 001

Latitude   Longitude

ASML   AGL

Tallest point on structure:

Type of cell:

Antenna	Sector	Orientation	Downtilt	Antenna type	ERP
	0				
	1				
	2				
	3				

401 contour required (Y/N) If yes attach contour

FAA study

Lighting/marking required (Y/N)

Colocated transmitters (Y/N)

  Call sign

  ERP

EMF power budget attached (Y/N)

Figure 4.8 FCC site information.

redesign might be in order. For example, in one actual case lowering of the antenna on a rooftop by 2 ft made the site FAA compatible. The overall key elements that need to be followed for compliance are:

1. Height

2. Glide slope

3. Alarming

4. Marking and lighting

Verification of height and glide slope calculations is needed for every site. It is recommended that every site have the FAA compliance checked and included in the master site reference document. If there is no documented record for a site regarding FAA compliance it is strongly recommended that one be made immediately. The time and effort required to check FAA compliance is not long; it could be done within a week for a several hundred cell system.

An example of a glide slope calculation is shown in Fig. 4.9. The actual glide slope calculation needs to be included with the site acceptance form to ensure that the process is done.

4.4 Planning and Zoning Board

Preparing for a zoning or planning board should be part of the design review process. Not only is the presentation important, it might be possible with a modification to the original site design to eliminate the process entirely. In many actual cases a modification to the site design would have eliminated the need to request a variance from the town and thus prevented massive delays in the cell site build pro-

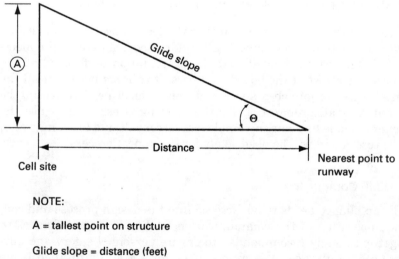

NOTE:

A = tallest point on structure

Glide slope = distance (feet)
                 factor

Factor = 100   if runway is > 3200 ft and within 20,000 ft of cell site
          50   if runway is < 3200 ft and within 10,000 ft of cell site
          25   if runway is a heliport and within 5000 ft of cell site

Figure 4.9 Glide slope calculation.

gram. While this step seems obvious, checking local ordinances and incorporating them into the design process is rare and more times than not a "forgotten child" in the design process.

It is recommended that the local ordinances for the site be included with the site source documents. However, when it comes time to present the case of why the site is needed to the local planning board, and often to the zoning board, a well-rehearsed presentation is needed. It is recommended that the program be rehearsed prior to the meeting night, to ensure that everyone knows what each one is going to say and when.

Engineering's role in the process tends to focus on why the site is needed and health and safety issues associated with electromagnetic fields (EMF). The items that should be presented or prepared for should include as a minimum:

1. Description of why the site is needed

2. Explanation of how the site will improve the network

3. Drawing of what the site will look like

4. Views from local residences

5. EMF compliance chart

6. EMF information sheets and handouts for the audience

Before the meeting it is essential that the local concerns be identified in advance so they can be specifically addressed before or at the meeting. It is also recommended that the presentation be focused at the public and not just at the board members. It is imperative that all the issues needed to launch a successful appeal for a negative ruling be covered. Overall preparation for the meeting is essential, since the comments made by company employees or consultants are a matter of public record and will be used solely for the appeal process.

4.5   EMF Compliance

EMF compliance needs to be factored into the design process and continued operation of the communication facility. The use of an EMF budget is strongly recommended to ensure personnel safety and government compliance. A simple source for EMF compliance issue should be the company's EMF policy.

The establishment of an EMF power budget should be incorporated into the master source documents for the site and stored on the site itself identifying the transmitters used, power, who calculated the numbers, and when it was last done. As a regular part of the preventive maintenance process the site should be checked for compliance and changes to the fundamental budget calculation.

The method for calculating the compliance issue is included in the IEEE C95.1-1991 specification with measurement techniques included in IEEE C95.3. A sample EMF power budget is included in Fig. 4.10. It should be signed off by the manager for the department and shared with the operations department of the company. An EMF budget needs to be completed for every cell site in operation and also for those proposed.

4.6 Frequency Planning

Frequency planning, or rather frequency management, is critical to the success of a communication system. Often it is the frequency planner who sets the direction for the performance of a communication system. The frequency planning process needs to be rigorously checked on a continuous basis to always refine the system. As a minimum for frequency planning, the designs, no matter how minor they seem, need to be passed through a design review process.

Frequency planning as a general rule is more of an art form than a defined science. The use of C/I ratios and whether you use an $N = 7$, $N = 4$, or $N = 12$ pattern are more clinical in nature. However, how you go about defining what the frequency plan design trade-offs are for an area is an art. Some engineers are better artists than others, but the fundamental issue of controlling interference and how well it is done falls on the design review process.

There are numerous technical books and articles, some of which are listed at the end of this chapter, that go over how to frequency plan a network, from a theoretical standpoint. It is very important to understand the fundamental principles of frequency planning in order to design a frequency management plan for any network. Failure to adhere to a defined frequency design guideline will limit the system's expansion capability.

The rationale behind defining this RF design process as an art is based on the multitude of perturbations available for any given frequency management plan. Several methods are available for use in defining the frequency management of a network. The method chosen by the mobile carrier needs to be factored into the frequency management scheme capacity requirements, capital outlays, and adjacent market integration issues, to mention a few. Obviously the method that is used for the frequency plan also has to ensure that the best possible C/I ratio is obtained for both cochannel and adjacent channel RF interference.

The use of a grid is essential for initial planning, but when "sprinkled with reality" the notion is academic in nature because the site acquisition process tends to drive the system configuration and not the other way around. Dealing with the irregularities of the site coverage,

Cell:					
Date:					

Sector 1:

Number of channels	19
ERP/channel	100 W
Total power	1900 W

Colocated transmitters:

Paging (931.875 MHz)	
ERP	1000 W

Data points:

Location 1	Distance	Total power	Power density	Max for band	% budget
Cell site	25 ft	1900 W	0.260527768	0.586666667	44%
Paging	20 ft	1000 W	0.214249809	0.62125	34%
				Total	79%

Location 2					
Cell site	100 ft	1900 W	0.016282986	0.586666667	3%
Paging	110 ft	1000 W	0.007082638	0.62125	1%
				Total	4%

Figure 4.10 EMF power budget.

traffic loading, and configurations requires continued maintenance of the network frequency plan. The specific channels available are defined based on the license you hold, either A- or B-band. In the United States an operator can operate only in either the A or B blocks and has available the spectrum range shown in Fig. 4.11. The spectrum chart shown is not contiguous, meaning other wireless operators are using parts of the spectrum adjacent to cellular receive and transmit.

Figure 4.12 is a simple frequency conversion chart which can be used to determine the exact frequency based on the FCC channel number used. As mentioned before, there are several different methods of assigning frequencies in a network. The most common methods utilized in cellular systems are $N = 12$, $N = 7$, and $N = 4$. Figures 4.13 to 4.15 will assist in laying out an initial plan using each of the methods mentioned.

The $N = 12$ method is usually deployed in an omni configuration system which is in areas that do not require the high-traffic-density carrying capacity provided in an $N = 7$ or $N = 4$ pattern. The $N = 12$ pattern is shown in Fig. 4.13. It is the most efficient for trunking efficiency since it is based on an omni cell site configuration.

The $N = 7$ method of channel assignments is one of the most popular methods for assigning frequencies. This pattern is shown in Fig. 4.14. There are several advantages with using an $N = 7$ pattern; they lie in its ability to provide a high level of trunking efficiency, increased flexibility for the placement of the RF channels in the net-

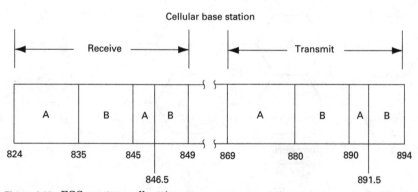

Figure 4.11 FCC spectrum allocation.

FCC channel (N)	Cell site transmit	Cell site receive
1–799	$(0.03)*N+870$ MHz	$(0.03)*N+825$ MHz
991–1023	$(0.03)*(N-1023)+870$ MHz	$(0.03)*(N-1023)+825$ MHz

Figure 4.12 FCC channel and frequency conversion chart.

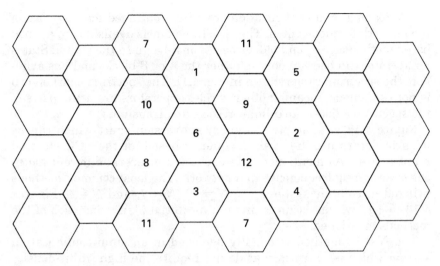

Figure 4.13 $N = 12$ frequency grid.

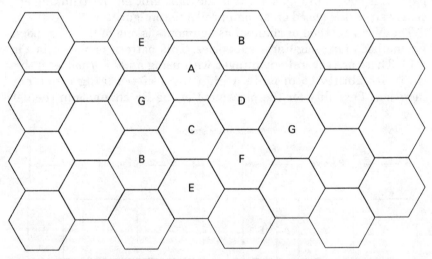

Figure 4.14 $N = 7$ frequency grid.

work, and excellent traffic capacity. The $N = 7$ pattern utilizes a 120° sectored cell design that equates to three sectors per cell site. This design facilitates the reuse of the same channels in close proximity to another cell site.

The $N = 4$ frequency-assignment method is another popular method for assigning frequencies. This pattern is shown in Fig. 4.15. The key advantage with using an $N = 4$ pattern is the high-traffic-capacity handling ability obtained. The $N = 4$ pattern delivers the most channels per square mile or kilometer of all the channel-assign-

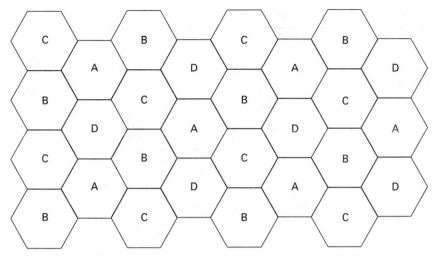

Figure 4.15 $N = 4$ frequency grid.

ment methods. However, the disadvantage is that this method is not as flexible for cell site placement and does not have the trunking efficiency under average traffic loads of the $N = 7$ pattern. The $N = 4$ pattern utilizes 60° sectors, six sectors per cell site, which accounts for its superior traffic-handling capabilities.

Most channel assignments utilize a 21-channel spacing rule, or 630 kHz between channels in a frequency group. The channel chart shown in Table 4.1 is typical for assigning frequencies at $N = 7$ reuse pattern. The same chart that is used for $N = 7$ is also used for $N = 12$ frequency assignments.

The channel chart used for an $N = 4$ pattern is similar to that used by an $N = 7$ pattern with just a few exceptions. In the $N = 4$ pattern 24 frequency groups are available before you exhaust the reuse pattern. With the $N = 7$ pattern 21 groups of channels are available; therefore, three more groups are required. The channel chart shown in Table 4.2 is used for an $N = 4$ pattern. The $N = 4$ channel chart, however, takes on a different flavor when using extended AMPS for the A-band owing to the channel spacing that physically takes place. The $N = 4$ channel chart shown needs to be modified to avoid intermodulation products when utilizing a duplexed antenna system.

Another method of channel assignment is used in the United States. It involves a 24-channel spacing frequency assignment chart. The 24-channel spacing chart shown in Table 4.3 reduces, or eliminates, adjacent channel interference in adjacent cell sites.

When looking at the various techniques employed for frequency assignments, it is obvious that there are many perturbations to the same objective. The frequency planning objective is to deliver the

TABLE 4.1   FCC Channel Chart for $N = 7$

Wireline B-Band Channels									
Channel group:	A1	B1	C1	D1	E1	F1	G1	A2	B2	C2
Control channel:	334	335	336	337	338	339	340	341	324	343
	355	356	357	358	359	360	361	362	345	364
	376	377	378	379	380	381	382	383	366	385
	397	398	399	400	401	402	403	404	387	406
	418	419	420	421	422	423	424	425	408	427
	439	440	441	442	443	444	445	446	429	448
	460	461	462	463	464	465	466	467	450	469
	481	482	483	484	485	486	487	488	471	490
	502	503	504	505	506	507	508	509	492	511
	523	524	525	526	527	528	529	530	513	532
	544	545	546	547	548	549	550	551	534	553
	565	566	567	568	569	570	571	572	555	574
	586	587	588	589	590	591	592	593	576	595
	607	608	609	610	611	612	613	614	597	616
	628	629	630	631	632	633	634	635	618	637
	649	650	651	652	653	654	655	656	639	658
	717	718	719	720	721	722	723	724	725	726
	738	739	740	741	742	743	744	745	746	747
	759	760	761	762	763	764	765	766	767	768
	780	781	782	783	784	785	786	787	788	789

Nonwireline A-Band Channels									
Channel group:	A1	B1	C1	D1	E1	F1	G1	A2	B2	C2
Control channel:	333	332	331	330	329	328	327	326	325	324
	312	311	310	309	308	307	306	305	304	303
	291	290	289	288	287	286	285	284	283	282
	270	269	268	267	266	265	264	263	262	261
	249	248	247	246	245	244	243	242	241	240
	228	227	226	225	224	223	222	221	220	219
	207	206	205	204	203	202	201	200	199	198
	186	185	184	183	182	181	180	179	178	177
	165	164	163	162	161	160	159	158	157	156
	144	143	142	141	140	139	138	137	136	135
	123	122	121	120	119	118	117	116	115	114
	102	101	100	99	98	97	96	95	94	93
	81	80	79	78	77	76	75	74	73	72
	60	59	58	57	56	55	54	53	52	51
	39	38	37	36	35	34	33	32	31	30
	18	17	16	15	14	13	12	11	10	9
	1020	1019	1018	1017	1016	1015	1014	1013	1012	1011
	999	998	997	996	995	994	993	992	991	716
	704	703	702	701	700	699	698	697	696	695
	683	682	681	680	679	678	677	676	675	674

TABLE 4.1   FCC Channel Chart for $N = 7$ (*Continued*)

Wireline B-Band Channels

D2	E2	F2	G2	A3	B3	C3	D3	E3	F3	G3
344	345	346	347	348	349	350	351	352	353	354
365	366	367	368	369	370	371	372	373	374	375
386	387	388	389	390	391	392	393	394	395	396
407	408	409	410	411	412	413	414	415	416	417
428	429	430	431	432	433	434	435	436	437	438
449	450	451	452	453	454	455	456	457	458	459
470	471	472	473	474	475	476	477	478	479	480
491	492	493	494	495	496	497	498	499	500	501
512	513	514	515	516	517	518	519	520	521	522
533	534	535	536	537	538	539	540	541	542	543
554	555	556	557	558	559	560	561	562	563	564
575	576	577	578	579	580	581	582	583	584	585
596	597	598	599	600	601	602	603	604	605	606
617	618	619	620	621	622	623	624	625	626	627
638	639	640	641	642	643	644	645	646	647	648
659	660	661	662	663	664	665	666			
727	728	729	730	731	732	733	734	735	736	737
748	749	750	751	752	753	754	755	756	757	758
769	770	771	772	773	774	775	776	777	778	779
790	791	792	793	794	795	796	797	798	799	

Nonwireline A-Band Channels

D2	E2	F2	G2	A3	B3	C3	D3	E3	F3	G3
323	322	321	320	319	318	317	316	315	314	313
302	301	300	299	298	297	296	295	294	293	292
281	280	279	278	277	276	275	274	273	272	271
260	259	258	257	256	255	254	253	252	251	250
239	238	237	236	235	234	233	232	231	230	229
218	217	216	215	214	213	212	211	210	209	208
197	196	195	194	193	192	191	190	189	188	187
176	175	174	173	172	171	170	169	168	167	166
155	154	153	152	151	150	149	148	147	146	145
134	133	132	131	130	129	128	127	126	125	124
113	112	111	110	109	108	107	106	105	104	103
92	91	90	89	88	87	86	85	84	83	82
71	70	69	68	67	66	65	64	63	62	61
50	49	48	47	46	45	44	43	42	41	40
29	28	27	26	25	24	23	22	21	20	19
8	7	6	5	4	3	2	1			
								1023	1022	1021
1010	1009	1008	1007	1006	1005	1004	1003	1002	1001	1000
715	714	713	712	711	710	709	708	707	706	705
694	693	692	691	690	689	688	687	686	685	684
673	672	671	670	669	668	667				

TABLE 4.2   FCC Channel Chart for $N = 4$

			Wireline B-Band Channels							
Channel group:	A1	B1	C1	D1	A2	B2	C2	D2	A3	B3
Control channel:	334	335	336	337	338	339	340	341	324	343
		356			359	360	361	362	345	364
		377			380	381	382	383	366	385
		398			401	402	403	404	387	406
		419			422	423	424	425	408	427
		440			443	444	445	446	429	448
	460	461	462	463		465			450	469
	481	482	483	484		486			471	490
	502	503	504	505		507			492	511
	523	524	525	526	527	528	529	530	531	532
	544	545	546	547	548	549	550	551		553
	565	566	567	568	569	570	571	572		574
	586	587	588	589	590	591	592	593		595
	607	608	609	610	611	612	613	614	615	616
	628	629	630	631	632	633	634	635	636	637
	649	650	651	652	653	654	655	656	657	658
	717	718	719	720	721	722	723	724	725	726
	741	742	743	744	745	746	747	748	749	750
	765	766	767	768	769	770	771	772	773	774
	789	790	791	792	793	794	795	796	797	798

			Nonwireline A-Band Channels							
Channel group:	A1	B1	C1	D1	A2	B2	C2	D2	A3	B3
Control channel:	333	332	331	330	329	328	327	326	325	324
	312	311	310	309	308	307	306	305	304	303
		290			287	286	285	284	283	282
		269			266	265	264	263	262	261
		248			245	244	243	242	241	240
		227			224	223	222	221	220	219
		206			203	202	201	200	199	198
	186	185	184	183		181			178	177
	165	164	163	162		160			157	156
	144	143	142	141		139			136	135
	123	122	121	120	119	118	117	116		114
	102	101	100	99	98	97	96	95		93
	81	80	79	78	77	76	75	74		72
	60	59	58	57	56	55	54	53	52	51
	39	38	37	36	35	34	33	32	31	30
	18	17	16	15	14	13	12	11	10	9
	1020	1019	1018	1017	1016	1015	1014	1013	1012	1011
	999	998	997	996		994			991	
	711	710	709	708	707	706	705	704	703	702
	690	689	688	687	686	685	684	683	682	681
	669	668	667							

TABLE 4.2   FCC Channel Chart for *N* = 4 (*Continued*)

Wireline B-Band Channels

C3	D3	A4	B4	C4	D4	A5	B5	C5	D5	B6	A6	C6	D6
344	345	346	347	348	349	350	351	352	353	354			
365	366	367	368	369	370	371	372	373	374	375	355	357	358
386	387	388	389	390	391	392	393	394	395	396	376	378	379
407	408	409	410	411	412	413	414	415	416	417	397	399	400
428	429	430	431	432	433	434	435	436	437	438	418	420	421
449	450	451	452	453	454	455	456	457	458	459	439	441	442
470	471	472	473	474	475	476	477	478	479	480	509	488	467
491	492	493	494	495	496	497	498	499	500	501	485	487	488
512	513	514	515	516	517	518	519	520	521	522	506	508	509
533	534	535	536	537	538	539	540	541	542	543	552	554	555
		556	557	558	559	560	561	562	563	564	573	575	576
		577	578	579	580	581	582	583	584	585	594	596	597
		598	599	600	601	602	603	604	605	606	623	625	626
617	618	619	620	621	622		624			627	644	646	647
638	639	640	641	642	643		645			648	665		
659	660	661	662	663	664		666						
727	728	729	730	731	732	733	734	735	736	737	738	739	740
751	752	753	754	755	756	757	758	759	760	761	762	763	764
775	776	777	778	779	780	781	782	783	784	785	786	787	788
799													

Nonwireline A-Band Channels

C3	D3	A4	B4	C4	D4	A5	B5	C5	D5	B6	A6	C6	D6
323	322	321	320	319	318	317	316	315	314	313			
302	301	300	299	298	297	296	295	294	293	292			
281	280	279	278	277	276	275	274	273	272	271	291	289	288
260	259	258	257	256	255	254	253	252	251	250	270	268	267
239	238	237	236	235	234	233	232	231	230	229	249	247	246
218	217	216	215	214	213	212	211	210	209	208	228	226	225
197	196	195	194	193	192	191	190	189	188	187	207	205	204
176	175	174	173	172	171	170	169	168	167	166	182	180	179
155	154	153	152	151	150	149	148	147	146	145	161	159	158
134	133	132	131	130	129	128	127	126	125	124	140	138	137
		111	110	109	108	107	106	105	104	103	115	113	112
		90	89	88	87	86	85	84	83	82	94	92	91
		69	68	67	66	65	64	63	62	61	73	71	70
50	49	48	47	46	45		43			40	44	42	41
29	28	27	26	25	24		22			19	23	21	20
8	7	6	5	4	3		1			2			
								1023	1022	1021			
1010	1009	1008	1007	1006	1005	1004	1003	1002	1001	1000	995	993	992
						716	715	714	713	712			
701	700	699	698	697	696	695	694	693	692	691			
680	679	678	677	676	675	674	673	672	671	670			

TABLE 4.3   FCC Channel Chart for N = 4 with 24 Channel Spacing

				Wireline B-Band Channels					
Channel Group:	A1	B1	C1	D1	A2	B2	C2	D2	A3	B3
Control channel:	334	335	336	337	338	339	340	341	342	343
	355	356	357	358	359	360	361	362	363	364
	379	380	381	382	383	384	385	386	387	388
	403	404	405	406	407	408	409	410	411	412
	427	428	429	430	431	432	433	434	435	436
	451	452	453	454	455	456	457	458	459	460
	475	476	477	478	479	480	481	482	483	484
	499	500	501	502	503	504	505	506	507	508
	523	524	525	526	527	528	529	530	531	532
	547	548	549	550	551	552	553	554	555	556
	571	572	573	574	575	576	577	578	579	580
	595	596	597	598	599	600	601	602	603	604
	619	620	621	622	623	624	625	626	627	628
	643	644	645	646	647	648	649	650	651	652
	717	718	719	720	721	722	723	724	725	726
	741	742	743	744	745	746	747	748	749	750
	765	766	767	768	769	770	771	772	773	774
	789	790	791	792	793	794	795	796	797	798

				Nonwireline A-Band Channels					
Channel group:	A1	B1	C1	D1	A2	B2	C2	D2	A3	B3
Control channel:	333	332	331	330	329	328	327	326	325	324
	312	311	310	309	308	307	306	305	304	303
	288	287	286	285	284	283	282	281	280	279
	264	263	262	261	260	259	258	257	256	255
	240	239	238	237	236	235	234	233	232	231
	216	215	214	213	212	211	210	209	208	207
	192	191	190	189	188	187	186	185	184	183
	168	167	166	165	164	163	162	161	160	159
	144	143	142	141	140	139	138	137	136	135
	120	119	118	117	116	115	114	113	112	111
	96	95	94	93	92	91	90	89	88	87
	72	71	70	69	68	67	66	65	64	63
	48	47	46	45	44	43	42	41	40	39
	24	23	22	21	20	19	18	17	16	15
	1023	1022	1021	1020	1019	1018	1017	1016	1015	1014
	999	998	997	996	995	994	993	992	991	716
	701	700	699	698	697	696	695	694	693	692
	677	676	675	674	673	672	671	670	669	668

TABLE 4.3   FCC Channel Chart for *N* = 4 with 24 Channel Spacing (*Continued*)

Wireline B-Band Channels													
C3	D3	A4	B4	C4	D4	A5	B5	C5	D5	B6	A6	C6	D6
344	345	346	347	348	349	350	351	352	353	354			
365	366	367	368	369	370	371	372	373	374	375	376	377	378
389	390	391	392	393	394	395	396	397	398	399	400	401	402
413	414	415	416	417	418	419	420	421	422	423	424	425	426
437	438	439	440	441	442	443	444	445	446	447	448	449	450
461	462	463	464	465	466	467	468	469	470	471	472	473	474
485	486	487	488	489	490	491	492	493	494	495	496	497	498
509	510	511	512	513	514	515	516	517	518	519	520	521	522
533	534	535	536	537	538	539	540	541	542	543	544	545	546
557	558	559	560	561	562	563	564	565	566	567	568	569	570
581	582	583	584	585	586	587	588	589	590	591	592	593	594
605	606	607	608	609	610	611	612	613	614	615	616	617	618
629	630	631	632	633	634	635	636	637	638	639	640	641	642
653	654	655	656	657	658	659	660	661	662	663	664	665	666
727	728	729	730	731	732	733	734	735	736	737	738	739	740
751	752	753	754	755	756	757	758	759	760	761	762	763	764
775	776	777	778	779	780	781	782	783	784	785	786	787	788
799													

Nonwireline A-Band Channels													
C3	D3	A4	B4	C4	D4	A5	B5	C5	D5	B6	A6	C6	D6
323	322	321	320	319	318	317	316	315	314	313			
302	301	300	299	298	297	296	295	294	293	292	291	290	289
278	277	276	275	274	273	272	271	270	269	268	267	266	265
254	253	252	251	250	249	248	247	246	245	244	243	242	241
230	229	228	227	226	225	224	223	222	221	220	219	218	217
206	205	204	203	202	201	200	199	198	197	196	195	194	193
182	181	180	179	178	177	176	175	174	173	172	171	170	169
158	157	156	155	154	153	152	151	150	149	148	147	146	145
134	133	132	131	130	129	128	127	126	125	124	123	122	121
110	109	108	107	106	105	104	103	102	101	100	99	98	97
86	85	84	83	82	81	80	79	78	77	76	75	74	73
62	61	60	59	58	57	56	55	54	53	52	51	50	49
38	37	36	35	34	33	32	31	30	29	28	27	26	25
14	13	12	11	10	9	8	7	6	5	4	3	2	1
1013	1012	1011	1010	1009	1008	1007	1006	1005	1004	1003	1002	1001	1000
715	714	713	712	711	710	709	708	707	706	705	704	703	702
691	690	689	688	687	686	685	684	683	682	681	680	679	678
667													

most radios to a given area with the least infrastructure and maintenance costs.

4.6.1 Basic guidelines

The main idea to keep in mind when frequency managing a system is
to maximize the distance between reusers; it's that simple a concept.
Whenever method or process is used when maximizing the reuse distance, it will be wrought with problems. When selecting a frequency
set for a cell site, whether it is a new site, expansion issue, or correcting an interference problem, the current and future configuration of
the network needs to be factored into the equation.

Many of us can recall the simple question, "who selected those frequencies and why?" It is recommended that a regular program of
retuning the network take place when adding sites and radios to the
network. The objective behind implementing regular retunes is to
help define a schedule whereby planning for channel and site expansion can take place. The use of regular retunes also enables a review
of the compromises implemented into the network and establishing a
plan for their correction in a regulated fashion. For example, you can
divide your system up into three or four sections and proceed in
either a clockwise or counterclockwise pattern for retuning the network. It is recommended that one region be retuned every 3 or 4
months depending on the size of the network.

Here are some basic guidelines that can and should be used when
putting together a frequency management plan for a network.

Rules for assigning voice channels

1. Do not assign cochannels or adjacent channels at the same cell site.

2. Do not assign cochannels in adjacent cell sites.

3. Do not mix and match frequency assignment groups in a cell or
   sector.

4. Avoid adjacent channel assignments in adjacent cell sites.

5. Maintain proper channel spacing, 630 kHz, in any channel assignments for a sector or site.

6. Maximize the distance between reusing cell sites.

Supervisor audio tone (SAT) assignment guidelines

1. Maximize the distance between reusing cochannel SAT sites.

2. Do not assign adjacent channels the same SAT.

3. Always examine handoff hub potentials when assigning SATs.

4. Avoid using multiple SATs at a cell site.

Control channel assignment guidelines

1. Do not assign cocontrol channels or adjacent control channels at the same site.

2. Do not assign cocontrol channels in adjacent cell sites.

3. Avoid having a control channel plan and a voice channel assignment plan using different reuse patterns.

4. Do not assign adjacent control channels in adjacent cell sites if they are directed at the same area.

5. Maintain proper channel spacing, 630 kHz, in any channel assignments of the sector and site.

Digital color code (DCC) assignment guidelines

1. Remember that four analog DCCs are available for assignment.

2. Maximize the distance between reusing cosetup DCC sites.

3. Always examine DCC assignments for dual origination problems.

4. Avoid using multiple DCCs at a cell site.

4.6.2 Frequency planning checklist

When putting together a frequency plan for a site or a system it is essential that a process exists that will ensure the major items are always reviewed. Since a large amount of artistic leeway is involved with frequency planning, i.e., more than one design can exist for a given situation, the use of a checklist will expedite the review process and ensure the accuracy of the plan.

The checklist below should be used when presenting any frequency plan. It accomplishes several key elements. First it forces the engineer to think about a multitude of items when putting the plan together and prevents common mistakes from occurring. Second it enables the reviewer and designer to identify any design issues not accounted for. Knowing what was not checked is essential in determining the degree of risk associated with a design.

There are several frequency assignment checklist items depending upon what the actual frequency management issue is. For instance, the frequency management activities associated with a new site, an existing site, or an area retune are similar though different. The checklists below are proposed lists which can and should be modified to reflect the particulars for your network.

New cell.   *Voice channel assignments*

1. Number of radio channels predicted
2. Cell site antenna orientation standard or nonstandard
3. Proposed ERP levels by sector
4. Coverage prediction plots generated
5. C/I prediction plots generated
6. Cochannel reusers for next three rings identified
7. Adjacent channel cell sites identified
8. SAT assignments checked for cochannel and adjacent channel
9. Link budget balance checked

Control channel assignments

1. Coverage prediction plots generated
2. Cochannel C/I plots generated
3. Proposed ERP levels by sector
4. DCC assignments checked for dual originations
5. 333/334 potential conflict checked

Frequency design reviewed by

1. RF design engineer
2. Performance engineer
3. Engineering managers

Existing sites frequency management checklist.   *Voice channel assignments*

1. Reason for change
2. Number of radio channels predicted
3. Cell site antenna orientation standard or nonstandard
4. Proposed ERP levels by sector
5. Coverage prediction plots generated
6. C/I prediction plots generated
7. Cochannel reusers for next three rings identified
8. Adjacent channel cell sites identified
9. SAT assignments checked for cochannel and adjacent channel

10. Link budget balance checked

Control channel assignments

1. Reason for change
2. Coverage prediction plots generated
3. Cochannel C/I plots generated
4. Proposed ERP levels by sector
5. Cochannel and adjacent control channels identified
6. DCC assignments checked for dual originations
7. 333/334 potential conflict checked

Frequency design reviewed by

1. RF design engineer
2. Performance engineer
3. Engineering managers

Area retunes. *Voice channel assignments*

1. Reason for change
2. Number of radio channels predicted for all sites
3. New sites expected to be added
4. Proposed ERP levels by sector for all sites
5. Coverage prediction plots generated
6. C/I prediction plots generated
7. Cochannel reusers identified by channel and SAT
8. Adjacent channel cell sites identified by channel and SAT
9. SAT assignments checked for cochannel and adjacent channel
10. Link budget balance checked

Control channel assignments

1. Reason for change
2. Coverage prediction plots generated
3. Cochannel C/I plots generated
4. Proposed ERP levels by sector
5. Cocontrol channel reusers identified by channel and DCC

6. Adjacent control channel reusers identified by channel

7. DCC assignments checked for dual originations

8. 333/334 potential conflict checked

Frequency design reviewed by

1. RF design engineer

2. Performance engineer

3. Engineering managers

4. Adjacent markets (if required)

The final requirement of any frequency plan proposal must include a written document issued to the engineering department (Fig. 4.16). It is recommended that the written document take on the form shown to ensure a paper trail is started and maintained. Every frequency change, channel assignment, SAT, and DCC alteration needs to have this document attached.

4.6.3 Test plans

When implementing a frequency plan it is essential that a test plan be developed. For example, actually altering the frequencies at a site usually requires assistance from the operations department. In the event that additional radios and potentially spans are needed there is

---

Document number:
Date:

Subject:

The following alterations are required to be implemented into the network:

Cell: Current: Future: Required date:

Design engineer _____

Engineering manager _____

Figure 4.16 Frequency assignment addition and change form.

another logistics piece that must be overcome. The test plan process or items identified below are used for a system or regional retune. When making individual channel or localized retunes a similar process needs to be followed but on a lesser scale.

The test plan generally needs to have several key elements, some of which are listed below.

1. Design objective

2. Design reviews

3. Coordination meeting and timelines

4. MOP

5. Implementation

6. Postanalysis

The design objective for the project can be crafted as a single paragraph which forms the basis of informing the company what the engineering department is trying to accomplish.

If the frequency plan involves a major system alteration, additional design reviews should take place with other departments within the company that are affected by this plan. It is essential in a major system retune that all the personnel in the engineering, operations, and implementation departments be involved with the process since they will be affected. The reviews should utilize the checklist as the basic meeting driver.

Once a design is agreed upon it will be necessary to call a general meeting to review the initial timetables focusing on the deliverables of each organization. The objective of this meeting is to identify where the critical paths are and if the schedule proposed is realistic or not. The coordination with adjacent markets might be an essential element for the timetable if they have to alter their system or if they have not reviewed the plan.

The next step that needs to be done is the generation of the method of procedure (MOP). The MOP will be the source document from this point forward identifying who does what and when. It is recommended that the MOP be as detailed as possible including an escalation procedure to be followed for notifying key company personnel of the project status. Every operator has an MOP, but a sample is included below for general reference. It needs to have a back-out procedure in the event that things go significantly wrong.

Figure 4.17 is a generic frequency planning MOP that can be used. It is highly advisable that you modify the MOP presented here to address specifics for the project at hand plus internal organization issues.

Preretune process

Date

X-X-XX Retune area defined
X-X-XX project leader(s) defined and timetables specified as well as
    the scope of work associated with the project
X-X-XX Traffic engineering provides radio channel count
X-X-XX Frequency planning begins design
X-X-XX Phase 1 design review (frequency planning only)
X-X-XX Phase 2 design review (all engineering)
X-X-XX Phase 3 design review (operations and engineering)
X-X-XX Phase 4 design review (adjacent markets if applicable)
X-X-XX Frequency assignment sheets given to operations
X-X-XX Retune integration procedure meeting
X-X-XX Executive decision to proceed with retune
X-X-XX Adjacent markets contacted and informed of decision
X-X-XX Secure postretune war room area
X-X-XX Briefing meeting with implementors of retune
X-X-XX Management information systems (MIS) support group confirms readiness
    for postprocessing efforts
X-X-XX Customer care and sales notified of impending actions

Retune process (begins X-X-XX at time XXXX)

X-X-XX Operations informs key personnel of retune results
    Operations personnel conduct brief postretune test to ensure call processing is
        working on every channel changed
    Operations manager notified key personnel of testing results

Postretune process (begins X-X-XX at time XXXX)

    Voice mail message left from engineering indicating status of retune (time)
    Begin postretune drive testing phase 1 (time)
    Datebase check takes place
    Statistics analysis takes place
    Voice mail message left from RF engineering indicating status of postretune
        effort (time)
    Phase 2 of postretune drive testing begins
    Commit decision made with directors for retune (time)
    Phase 3 of postretune drive testing begins

X-X-XX

    Continue drive testing network
    Statistics analysis
    Conduct postretune analysis and corrections where required

X-X-XX

    Postretune closure report produced

Figure 4.17   Method of procedure for regional retune.

The implementation of the plan is probably the most exciting and nerve-racking part of the process because this is where the true design review takes place. The actual implementation of the plan is covered in the MOP and in all likelihood is carried out by the operations department. How you implement the program is dependent upon whether you have manual or autotuning combiners or LACs. Several approaches have been done, ranging from a phased approach of altering one channel set at a time to flash cutting the network and letting the interference fall where it will during the transition process. The particular approach is dependent upon the design and logistics factors involved with the process.

Identification of the drive routes that will be utilized as part of the postretune process is essential and should be included in the design review and implementation process. The drive routes should be sectioned into three categories. Category 1 should be the critical location where a possible problem would be most likely to occur. The location 1 spots need to be focused on immediately after the retune takes place. Location 2 are all the other roads that need to be part of the process. Location 3 roads and routes are those that arise as a result of the previous postanalysis phases 1 and 2 where problems are identified and restoration action takes place.

Postanalysis of the retune is one area that allows all involved to know if their efforts were successful or not. When conducting a retune it is essential that a final report comparing pre- and postchange data is completed. It is also recommended that for the first week after the retune a daily status report be issued identifying all the problems found and the fixes implemented to date. No retune takes place that does not have some level of problems associated with it, so it is best to plan for these with proper documentation.

The first week's data should be monitored on an hourly basis and then relaxed to a daily basis. The data should be compared with the previous week's data, day by day and hour by hour, for comparison. The small amount of data, i.e., time frame, will tend to make things confusing but will identify the worst sites in the network and expedite the troubleshooting process.

The data reviewed should include the following items:

1. Lost call

2. Blocked or sealed channels

3. Access failures

4. Out-of-service channels (OOS) channels

5. Customer complaints

6. Drive test data

7. Usage/lost calls

8. Periodic locate measurements (PLMs) or equivalent

Finally, since the data extracted are most likely to come from various support systems in the network, it is essential that the MIS know about the system retune to prepare any data required by the engineering department and to be aware of the possible changes in other company reports.

4.7 Radio Expansion

Radio expansion is an ongoing process of a growing wireless communication system. There is always the need to determine how much blocking in the system is acceptable. Radio expansions should take place on a 6-month basis for all the sites in the network. The rationale behind utilizing a 6-month time frame is that this schedule allows for better utilization of company capital equipment and minimizing the ongoing expense of leased T1 lines for cell sites and port expanse with switches.

The frequency plan should incorporate a yearly projection of additional voice channels for planning purposes and should correspond with a rotational retune program. However, for better channel management any new channels should be implemented in a logical fashion with the operations department 1 month before they are needed by the system.

The guidelines to follow for this effort involve the monitoring of the voice channel blocking levels, percent utilization, capital, and lease line expenses. The first step that must be followed is to determine the current system growth rate that is taking place. These data can be secured from marketing, or you can do a linear approximation following system growth. One caveat that needs to be introduced here is that the system traffic growth is seasonal and should be incorporated into the modeling data, but conducting a 1-year projection almost negates this approach.

The issue of peak versus average traffic needs to be resolved before the system growth projection can be completed. One method that works involves utilizing an average of the 10 busiest days of the month for the peak month in that year during the system busy hour. This method has been used with much success and will remove peak traffic spikes from overinflating your capital deployment of radios. The seasonal traffic adjustments can be used to facilitate frequency planning by minimizing the number of reused channels in the network. The benefit here is reduced overall network interference.

The utilization rate of the voice channels used in a network can range from 80 to 100 percent depending upon budgetary and marketing requirements. It is recommended that the 80 percent utilization level be used as the trigger point for determining when additional radios will need to be placed in service in the network.

The blocking level used for a network tends to vary according to an operator's design philosophy. Some standard blocking levels range from a maximum of 2 percent and a minimum of 1 percent. The use of the 2 percent maximum follows traditional cellular engineering and may be downward adjusted for competitive reasons. The type of blocking table used should be checked and noted. Most companies utilize an erlang B table while some make use of a poisson table. Be aware that a few traffic charts are available that promote themselves as erlang B but are in fact not that but some hybrid approach that follows no logical pattern except to add channels.

The following is a brief radio expansion procedure to be used as a guide for what to do when adding radios to the system. The process below can be led by either the manager for network or performance engineering.

On a 6-month basis:

1. Determine, based on growth levels, the system needs on a 1-, 3-, 6-, and 12-month period.

2. Factor into the process expected sites available from the build program, accounting for deloading issues.

3. Establish the number of radios that need to be added or removed from each sector in the network.

4. Modify the quarterly plan previously issued.

5. Determine the facilities and equipment bays needs to support channel expansions.

6. Inform the frequency planners, performance engineers, equipment engineers, and operations personnel of their requirements.

7. Issue a tracking report showing the status of the sites requiring action. This report should contain as a minimum the following items:

- Radios currently at every site
- Net change in radios
- When exhaustion, 2 percent blocking, is expected to occur
- Radios ordered (if needed)
- Facilities ordered (if required)

- Radios secured
- Frequencies issued
- Cell site translations completed
- Facilities secured (if needed)
- Radios installed or removed
- Activation date planned for radios

8. On a quarterly basis conduct a brief (1-hour) meeting to discuss the provisioning requirements and arrangements for personnel.

9. Perform a biweekly traffic analysis report to validate the quarterly plan and issue the status report of the progress at the same time.

The last step in the system tracking process involves the activation date planned for the radios. This particular piece will enable you to preposition the radios in the network without actually activating them. Specifically if the radios are needed for the midpoint of March but are installed in January you can keep the interference levels to a minimum by having them remain out of service. The postimplementation of the channels and the task of tracking what stage you are in is essential and included as part of the checklist above. It is recommended that a MOP be generated and followed to ensure that the radios are properly added into the network.

4.8 Antenna Change or Alteration

The alteration of an antenna system, whether it is orientation, degree of inclination, or change of antenna type can have a large impact both positive and negative upon a network. Ensuring that any antenna system change is done with a design review process is essential to the health of a communication network.

The first step in the process is to define just what the purpose is for this change and the benefits expected. Often this simple step is overlooked and a change takes place with many capital dollars being expended without anyone's truly knowing the desired end result. The criteria used for defining the antenna system change need to be well defined. For example, if the intent is to improve coverage by increasing the gain of the antennae a link budget analysis needs to be performed. If the desire is to utilize a narrower horizontal antenna pattern, the impact to the current coverage in the area handled by the site needs to be evaluated as a minimum. A third example would be for downtilting of the antenna where the objective would be to reduce interference and/or current cell coverage and alternative methods of achieving the same result should be investigated.

As with all antenna changes the internal FCC process needs to be followed to ensure the proper documentation is maintained and the site source documents are updated. After a design change has been completed it is essential that some form of posttesting take place to ensure that the cure is not more problematic than the illness you were trying to correct. Pre- and posttesting is essential for closing the design loop process. Drive testing of the affected area before and after these changes needs to take place.

Statistical analysis will also need to be done to ensure that adjacent sites are not adversely affected. The following is a simple checklist for important metrics that need to be reviewed:

1. Reasons for changing
2. Criteria for selecting antenna system
3. Desired results
4. Pretesting plan
5. Implementation plan
6. Internal and external coordination
7. FCC internal process
8. Posttesting
9. Conclusion report
10. Cell site source file update

4.9 Parameter Settings and Adjustments

The alteration of existing site software parameters is often the least documented facet of the ongoing operation of a network. The need to ensure that a paper trail is associated with every site is essential to expedite troubleshooting and prevent old problems from reappearing with new engineers being assigned to the site. As always the first step in the process is to define just what the objective is for this change and the benefits expected. Often this simple step is overlooked and a change takes place, with the proposed solution never being achieved.

The criteria used for defining software parameter changes need to be well defined. For example, if the desire is to shed traffic from a particular site then the adjacent site, which most likely will receive the overflow of traffic, needs to be checked for sufficient capacity. If the intent is to change the handoff table for a given site then a check of the frequency plan for potential handoff problems needs to be completed.

Statistical analysis will also need to be done to ensure that adjacent sites are not adversely affected. The simple checklist for what to look for is given below.

1. Reasons for changing

2. Criteria for parameter change

3. Desired results

4. Pretesting plan

5. Implementation plan

6. Internal and external coordination

7. Posttesting

8. Conclusion report

9. Cell site source file update

Although the list is elementary, it should be followed as a minimum.

References

1. American Radio Relay League, *The ARRL 1986 Handbook,* 63d ed., The American Radio Relay League, Newington, Conn., 1986.
2. *The ARRL Antenna Book,* 14th ed., The American Radio Relay League, 1984.
3. AT&T, *Engineering and Operations in the Bell System,* 2d ed., AT&T Bell Laboratories, Murray Hill, N.J., 1983.
4. Carlson, A. B., *Communication Systems,* 2d ed., McGraw-Hill, New York, 1975.
5. Carr, J. J., *Practical Antenna Handbook,* TAB Books, Blue Ridge Summit, Pa., 1989.
6. Code of Federal Regulations, CFR 47 parts 1, 22, and 24.
7. DeGarmo, Canada, Sullivan, *Engineering Economy,* 6th ed., Macmillan, New York, 1979.
8. Dixon, *Spread Spectrum Systems,* 2d ed., Wiley, 1984.
9. Hess, *Land-Mobile Radio System Engineering,* Artech House, Norwood, Mass., 1993.
10. IEEE C95.1-1991, IEEE Standard for Safety Levels with Respect to Human Exposure to Radio Frequency Electromagnetic Fields, 3 kHz to 300 GHz.
11. ITT, *Reference Data for Radio Engineers,* 6th ed., Howard W. Sams & Co., New York, 1983.
12. Jakes, W. C., *Microwave Mobile Communications,* IEEE Press, New York, 1974.
13. Johnson, R. C., and H. Jasik, *Antenna Engineering Handbook,* 2d ed., McGraw-Hill, New York, 1984.
14. Kaufman, M., and A. H. Seidman, *Handbook of Electronics Calculations,* 2d ed., McGraw-Hill, New York, 1988.
15. Lathi, *Modern Digital and Analog Communication Systems,* CBS College Printing, New York, 1983.
16. Lee, W. C. Y., *Mobile Cellular Telecommunications Systems,* McGraw-Hill, New York, 1989.
17. Lee, W. C., *CoChannel Interference Reduction by Using a Notch in Tilted Antenna Pattern,* IEEE, New York, 1985.
18. Lee, W. C., *Effects on Correlation between Two Mobile Radio Base-Station Antennas,* IEEE, New York, 1972.
19. Yarbrough, *Electrical Engineering Reference Manual,* 5th ed., Professional Publications, Inc., Belmont, Calif., 1990.

Network Design Guidelines

This chapter covers the basic design guidelines used to build and operate a new cellular system. Further reading is advised on each of these topics to assist the network engineer in the development and design of a practicable, reliable, and efficient network design. Consult the references listed for this chapter for additional resources.

5.1 Specific Required System Design Parameters

Table 5.1 lists the parameters that are valuable in the design and operation of a cellular system. These data should be updated on a regular basis as input to the quarterly network design review and should be compared to the previous quarter to detect large changes in any specific parameter for determining if the modeling and prediction efforts by the network group are improving or remaining consistent.

5.2 System Switching Design

The following discussion provides some basic design considerations to take into account when choosing a switching product for your network.

5.2.1 System switch capacity (traffic)

The determination of the switching traffic capacity required for a new system is dependent upon the initial estimated traffic to be processed at the time of the system cut and for a specified period afterward, 1

TABLE 5.1   System Design Parameters

- Size of initial subscriber base
- Projected growth of subscriber base over a 2-year period
- Estimated usage per subscriber (millierlangs per subscriber)
- Estimated calls per subscriber
- Estimated calls per second
- Estimated average call holding time (data taken from other known systems in the area or a typical industry value used)
- Estimated switch initial traffic (erlangs)
- Projected switch traffic growth over a 2-year period
- Estimated switch calls processed per second
- Projected switch calls processed per second over a 2-year period
- Estimated number of cell sites in service at the time of system cut at the end of the specified design period
- Projected cell site growth over a specified design period
- Estimated number of PSTN trunks in service at the time of the system cut or at the end of the specified design period
- Projected number of PSTN trunks required over a 2-year period
- Estimated number of IMT trunks in service at the time of the system cut
- Estimated number of data links (SS7) in service at the time of the system cut
- Projected number of data links (SS7) required over a 2-year period
- Auxiliary systems in service at the time of system cut or at the time of the system design reviews:

Voice mail system (initial and projected required capacity)

Customer service department system interface to the network

Validation systems (precall and or postcall)

Fraud systems

Billing systems

Network management system

year, 2 years, etc. This traffic is estimated by the number of subscribers to be served for a designated coverage area at an estimated level of usage per subscriber. The traffic capacity of a switch itself is given as the maximum amount of traffic measured either in calls processed per second, erlangs (a dimensionless measurement based on the erlang B formula; see Sec. 6.4 for details), or CCS (100 call seconds per hour) that it can process within an acceptable level of processor loading.

Therefore, for an example system design that requires switching capacity to serve 50,000 subscribers at an estimated traffic level of

0.015 erlangs per subscriber and a 2-year projected subscriber growth of an additional 100,000 subscribers the switching capacity required is as follows. The initial traffic load would be 750 erlangs [estimated number of subscribers in the system (50,000) times the average usage or traffic level per subscriber (0.015) = 750 erlangs]. The 2-year projection would be 2250 erlangs [projected 2-year subscriber base growth (100,000) times the average usage or traffic level per subscriber (0.015) = 2500 erlangs]. A typical switch may have a capacity of 3000 to 5000 erlangs, which indicates that the switching capacity of the system would be met during the initial 2-year period. However, these values are highly dependent upon the application environment of the switch, and such projections need to be updated on a regular basis to constantly improve their accuracy. Table 5.2 presents the above data in a tabular form.

Capacity based upon the ability to process a certain number of calls per second is another metric used to determine the overall switching requirements of a system and to measure the performance of the switch itself. This measurement, however, is more complex than using the basic traffic load metric mentioned above because it takes into account the various components of a mobile phone call and their cumulative effect on the processor load of the switch. A mobile phone call can have many more functional pieces than a normal landline phone call. In a mobile environment a phone call can be of a land to mobile call type (L-M), a mobile to land call type (M-L), or a mobile to mobile call type (M-M). A mobile call could also include an intraswitch (within the same switch) handoff or an interswitch handoff (between two switches). The mobile call may involve the use of one of the many system features such as call waiting, call forwarding, or three-party conference. All these components that make up a mobile phone call will contribute to the processor load of the switch in varying amounts according to the environment in which the system will operate.

When a system has been in operation for a period of time the amounts and percentages of each of these call components can be

TABLE 5.2   Example System Design Data

Initial or current size of subscriber base	50,000 subscribers
2-year projected subscriber base growth	100,000 subscribers
Estimated or actual traffic level per subscriber	0.015 erlangs
Initial or current system traffic load	750 erlangs
2-year projected system traffic load	2250 erlangs
Normal switch traffic load (as recommended by the switch vendor)	3000–5000 erlangs

obtained along with the total number of calls processed during the busy hour. Using this metric to determine the switching requirements of an initial system will depend entirely upon estimated calling patterns of the projected subscriber base. I would recommend the use of this metric after the system has been in operation for a period of time. At this point it can be used for building a more accurate model of the traffic load on the switch and thus can better predict future switching requirements of the network.

As an example, let's assume a system has been in operation for a period of time (6 months). The distribution of call types, handoff percentages, and feature usage can now be obtained from the system statistics or from the billing reports generated by the information systems department. Once these data are available the next step is to determine what the switch processor utilization is for each of these functions individually. This can be obtained from the switch vendor, since they have the ability to measure the individual effect of each of these functions on the CPU load of the switch within a laboratory environment. With all these data available, a model of the switch CPU loads for the network can be assembled. Table 5.3 provides a simple model and switching requirement prediction example.

Switch CPU load calculation

$$\text{Switch CPU load} = 100\% - \text{CPU baseline load} - 100*[(0.009*A) + (0.013*B) + (0.0025*C) + (0.0015*D)]$$

$$= 100\% - 15\% - 27.2\% = 57.8\%$$

where switch CPU baseline load = 15%

$$A = \text{M-L} + \text{L-M} + 2(\text{M-M}) \text{ call attempts per second during system BH}$$

$$= 11.1 + 2.1 + 1.4 = 14.6$$

$$B = \text{M-L} + \text{L-M} + \text{M-M call completions per second during system BH}$$

$$= 7.8 + 1.0 + 0.3 = 9.1$$

$$C = \text{Intersystem HOs attempts per second during system BH}$$

$$= 8.3$$

$$D = \text{Intersystem HOs completions per second during system BH}$$

$$= 1.3$$

TABLE 5.3   Example Switch Capacity Model Using Call Components

System Data (mobile = M, land = L, system busy hour = BH, handoff = HO)	
Number of M-L call attempts during BH	40,000
Number of M-L call attempts per second during BH	11.1
Number of L-M call attempts during BH	7,500
Number of L-M call attempts per second during BH	2.1
Number of M-M call attempts during BH	2,500
Number of M-M call attempts per second during BH	0.7
Number of M-L call completions during BH	28,000
Number of M-L call completions per second during BH	7.8
Number of L-M call completions during BH	3,750
Number of L-M call completions per second during BH	1.0
Number of M-M call completions during BH	1,000
Number of M-M call completions per second during BH	0.3
Number of intrasystem HOs attempted during BH	120,000
Number of intrasystem HOs completed during BH	24,000
Number of intersystem HOs attempted during BH	30,000
Number of intersystem HOs attempted per second during BH	8.3
Number of intersystem HOs completed during BH	4,500
Number of intersystem HOs completed per second during BH	1.3
Percentage of M-L system calls	80%
Percentage of L-M system calls	15%
Percentage of M-M system calls	5%
M-L call completion rate	70%
L-M call completion rate	50%
M-M call completion rate	40%
HO completion rate	20%
Vendor-Supplied Data	
CPU utilization for M-L type call attempt per second	0.20%
CPU utilization for L-M type call attempt per second	0.30%
CPU utilization for M-M type call attempt per second	0.40%
CPU utilization for M-L type call completion per second	0.33%
CPU utilization for L-M type call completion per second	0.47%
CPU utilization for M-M type call completion per second	0.50%
CPU utilization for one intersystem HO attempt per second	0.25%
CPU utilization for one intersystem HO completion per second	0.15%

Take note, this is only an example model. The model supplied by your switch vendor will be tailored to the specific design of their switching product and the type of processor in the node itself, i.e., CPU, secondary processor, etc.

System switch capacity summary

1. For systems not yet in operation determine the system switch capacity by use of the total estimated traffic to be generated by the system in erlangs or CCS.

2. For systems currently in operation determine the system switch capacity after using the more accurate call distribution model.

5.2.2 System switch capacity (ports)

Estimating the number of ports required by the system at the initial in-service date will be a factor in determining the number of switches needed and the remaining port capacity of the network. In this book the term "switch port" or just "port" refers to an actual voice circuit within the switch itself. Every switch has a limited number of these ports available for use in the assignment and switching of both system RF voice circuits (channels) and landline voice circuits.

The number of switch ports needed for the initial system RF channel requirements is equal to the number of cell sites expected to be in service at the time of cut (turning on the system), and the number of voice channels assigned to each cell site will be assigned. These values are dependent upon the coverage area of the system, the expected size of the subscriber base, and the agreed-upon grade of service (GOS) the system is to provide. This input can be obtained from the RF engineering department or the designated group responsible for the initial design of the system.

The number of switch ports needed for system interconnection to the PSTN and between-system nodes must be determined and used as part of the total initial port requirements of the system. These values are dependent upon the estimated amount of traffic to be delivered between the mobile system, the PSTN, and the individual switching nodes of a multinode network at a specified GOS. The estimated traffic to be delivered between the mobile system and the PSTN can be further broken down into local (LEC) and long-distance or toll (IXC) calls. It would be expected to overdesign a system initially since the exact traffic patterns are not known. A 5 percent overcapacity is a reasonable figure to use as a guideline. However, once these patterns and volumes are measured the system network design can be adjusted for greater efficiency and flexibility.

From an RF perspective an example system design requiring 50 cell sites at 30 channels per site would require 1500 switch ports for the base system design (Table 5.4). The network group may require 2100 switch ports with a breakdown of 1200 ports for local traffic, 600 ports for toll traffic, and 300 ports for internal system traffic. Combined the two inputs give a total of 3600 ports as the initial system design requirement.

5.2.3 System switch capacity (subscriber storage)

The ability of a switch to store subscriber profile records is probably more important to a small system than to a large system. This is

TABLE 5.4   **Example Switch Port Capacity Model**

System Data (Cell Site Requirements)	
Number of expected cell sites to be in service	50
Number of RF channels estimated per cell site	30
Number of switch ports required for cell sites	1500
System Data (Network Requirements)	
Number of system local circuits	1200
Number of system toll circuits	600
Number of internal system circuits	300
Number of switch ports required for network	2100
Total system switch port requirements	3600

because a large system will have the ability to develop their network to a level where HLRs (dedicated system nodes mainly used to store subscriber records) have been implemented and the majority of the subscriber base loaded into these nodes. For a smaller system that cannot afford the expense and engineering resources to complete this network development the storage capacity of a switch will be critical to its operation.

An example system may have an expected subscriber base of 3000 to 5000 customers with a growing to projected year-end count of 50,000. A typical switch may have a subscriber storage capacity of 70,000 records.

5.2.4   System switch standardization, flexibility, and support

When choosing a switch product and design for an initial system the main factors, as listed above, are the traffic capacity, the port capacity, and the subscriber record storage capacity. However, there are other important considerations that need to be factored into the overall decision. A partial list and description of some of these is given below:

1. *Compatibility with industry standard telecommunication protocols.*   Is the switching platform compatible with the standard protocols used by the industry for a given area of application (i.e., domestic and international applications)? The ability for a switch to have signaling system 7 (SS7) capabilities for use in the United States is critical since this is the predominant networking protocol for switch-to-switch communications between differing vendor types.

2. *Standardized billing formats.*   Does the switch product support a standard record format for its billing structure? This is important,

especially when the operator is planning to use an outside billing vendor for processing customer accounts. When performing system software upgrades, does the vendor specify the exact changes (if any) in the billing record format prior to the new release being cut into service? Does the vendor continue to support the existing and new billing standards?

3. *Industry- and market-specific switch functions.* The switching product chosen should provide functions standard to the cellular industry plus any additional features the specific market may require. An example of both types of functions is listed below.

Standard switching functions:

- Subscriber features: three-party conference, call waiting, call forwarding, etc.
- Paging area definitions; have the ability to define multiple paging areas in a single switch
- Mobile autonomous registration; have the ability to perform automatic registration for compliant mobiles in the system
- Subscriber validation and denial functionality; have the ability to provide basic and simple mobile validation and service denial functions

Specialized switching functions:

- Specific support of a networking protocol
- Specialized paging functions for RF border applications
- Immediate and user-friendly mobile call trace functionality
- Real-time billing
- User friendly interface available to operational personnel

4. *Switch reliability and maintenance.* The switching product chosen should have an acceptable level of reliability as set by industry standards. The actual product reliability measurements can differ from vendor to vendor. Take the time to compare these specifications against one another for a better understanding of the methods used in evaluating this aspect of the switching product. The vendor may be able to supply actual generic reports with real performance data from active systems. This too would be helpful in selecting a switching product. In addition to reliability switch maintenance functions should be considered when selecting a switch. How difficult are these functions to perform? The more complex the product the more training required by the operations per-

sonnel. Does the switch use standardized command formats? How long does it take to complete these functions? Obviously the longer these maintenance tasks take to complete the larger the maintenance window for the system and the more time required by the operations personnel. Ultimately, this can mean larger labor costs to the company.

5. *Switch vendor software, hardware, and operations support.* What level of support does the switch vendor provide? A 24-hour support center with trained personnel? Is there a formal process with delivery schedules for development to complete system software and hardware fixes and future designs? Is there an exceptionally long delivery time for these items? Is development support available on site (at the customer's switch location) for the delivery and implementation of these items?

5.3   System Interconnect Design (Voice)

Cellular networks consist of both voice- and data-type communications. For the sake of clarity these two networks are separated into two different discussions. Some of the main topics in the design of a voice network are considered in this section while issues pertaining to the data network are given in the following section. In actual cellular systems voice- and data-type communications may be intertwined at various points in the network and isolated at others. Knowing and understanding the distinction between these types of communications is vital to the successful work of the network engineer.

5.3.1   Network facility types

Types of network facilities required are determined by reviewing the cellular services to be initially offered as proposed in the company business and marketing plan. For instance, if mobile customers will have the option to make international calls then facilities must be purchased between the cellular system and an area interexchange carrier (IXC) for providing this type of service. It will be necessary to contact the various carriers for the type, quantity, and cost of the interconnection facilities needed. These facility types are also determined by the type of interface used and whether it is supported by the network switch. Some example interface parameters are the use of MF or DTMF signaling, the type of trunk selection algorithm used, the type of glare algorithm used, etc. These and other parameter settings combined define the interface between the mobile switch and the LEC switches, the IXC switches, and the other intersystem switches in a multiple-node network. (See Table 5.5 for an example system interconnection plan.)

5.3.2 Network grade of service (GOS)

In the ideal network all voice-carrying facilities would be available 100 percent of the time and there would be no blocking of any call traffic. However, as nice as this sounds, the facility costs would be too great to design such a network. Any company attempting to do so would probably go bankrupt. Therefore, a level of acceptable blocking must be set and the network built around this value to obtain a practicable design. As mentioned in Sec. 4.7, by using the erlang B formula or referencing the erlang B table an engineer can find the required number of facilities for a given traffic load and grade of service. This method is easy and used widely throughout the telecommunications industry. Be aware that the grade of service for network facilities is typically better (lower blocking probability value) than the level set for the RF portion of a cellular system. Some typical values are $P = 0.001$ GOS for the network (land-type) facilities and $P = 0.02$ GOS for the RF system channels. The definition of the erlang is given in Sec. 6.5, and the use of the erlang B formula or table is shown in Sec. 9.2. Again, the basic idea is that, for a given grade of service and traffic level, an engineer can obtain the required number of facilities to provide the desired level of service.

As mentioned above in the switch port discussion, the expected traffic levels for local calls (LEC facilities), toll calls (IXC facilities), intrasystem calls (IMT facilities), and calls to any interconnected systems must be approximated to the best degree possible for determining the port requirements of the network switches and the number of network interconnect facilities needed. Since a newly designed system will not have actual data available for this, modeling data from other networks and general estimations must be used. In addition to the traffic level estimates and the selected grade of service, the facilities needed to provide alternate routing within the network must also be taken into consideration for a complete network design. These alternate route facilities will increase the total quantity of circuits required by the initial system design.

5.3.3 Network timing

The network timing can be derived from one of the central offices of the PSTN using a dedicated channel (time slot) of an interconnecting T-span as long as the source is a Stratum-2 level or greater. It is recommended that two such sources be available from the PSTN for net-

TABLE 5.5 Example System Interconnect Plan

Trunk group	Route	No. of Circuits	Facility type	Signaling type	Address	Comments
				PSTN (LEC) Facilities		
100	Route 001	4 T-spans	LEC	MF	Switch 1 to central office A	Note, span 1 of trunk group 100 is used for network timing.
101	Route 002	2 T-spans	LEC	MF	Switch 2 to central office B	
				PSTN (IXC) Facilities		
110	Route 003	4 T-spans	IXC	MF	Switch 1 to tandem office A	
112	Route 004	2 T-spans	IXC	MF	Switch 2 to tandem office B	
				Intermachine Trunks (IMTIs) (Internal System)		
115	Route 005	2 T-spans	IMT	MF	Switch 1 to switch 2	
116	Route 006	2 T-spans	IMT	MF	Switch 1 to switch 2	
				Intermachine Trunks (IMT's) (Outer System)		
120	Route 007	2 T-spans	IMT	MF	Switch 1 to system C	
121	Route 008	2 T-spans	IMT	MF	Switch 2 to system D	

work timing purposes. These sources are in addition to the backup timing source that most switches provide. The backup switch sources are usually a Stratum-3 level and thus are not accurate for use in a live network for normal operation (see Sec. 7.1 for more details on Stratum accuracy levels). Some companies may decide to buy their own Stratum-2 level source to fulfill the timing requirements of their specific network. The cost of these units has decreased considerably over the past years, enough to make this design choice a viable option. Take note, all network facilities used for timing purposes must be clearly identified with proper labeling and documentation to prevent them from being worked on without proper preparation and notification to both the engineering and operations departments.

5.3.4 Network diversity

The alternate routing provided in the network will be dependent upon many different factors: the type of facilities and equipment available for use; the number, location, and distances of the area serving central offices; the number of nodes in the cellular system along with their method of interconnection; budgetary constraints; etc. Typically a route in the network will have at least one alternate or backup route to serve as standby in the event the first route is out of service. More than one alternate route may be available for a given facility depending on the network design and the amount of redundancy required in the network. For instance, if microwave links are available and economical to the company to deploy then perhaps two alternate routes may be used between MTSOs for added redundancy and error recovery.

An example system alternate routing diagram is given in Fig. 5.1. Switch A routes both voice and data traffic to switch B using route A-B. Should the facilities carrying traffic for this route fail, then switch A would utilize route A-C to redirect this traffic to switch C. Switch C would then route this traffic destined for switch B over route C-B.

Summary of system interconnect design (voice)

1. Determine the types of interconnection facilities needed from the proposed service offering and the switch interface chosen.

2. Determine the grade of service to be used in the network design for landline-type facilities. Typical value is $P = 0.001$.

3. Determine the amount of diversity required in the network alternate routing design. A typical routing design will have at least one backup or secondary route per every primary route.

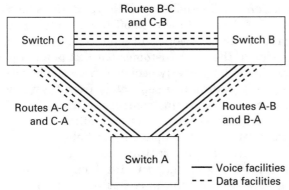

Figure 5.1   Example system alternate routing diagram.

4. Collect data for the network design, type of interconnection facilities available for use, the number, location, and distance of the area serving central offices for PSTN-type interconnections, etc.

5. Formulate an interconnection plan using the information from the previous steps to develop the recommended system tables and diagrams.

5.4   System Interconnect Design (Data)

The decision as to the type and design of the data network will again be based upon a number of factors. One of the most important aspects of this decision process is to determine what industry standard to follow for a particular application. For example, what type of physical interconnect standard should be deployed in the construction of the SS7 data network? Should the company use the V.35 standard or the OCU standard? Both specifications have their advantages and disadvantages that need to be taken into account depending on the operating environment and cost factors involved with the project.

These decisions are becoming more difficult in the ever-changing world of telecommunications. Other considerations involve the individual parameters of the application. When discussing data networks, the main emphasis of this book is on signaling system number 7 (SS7) since this is the most widely used protocol for data message transport between switches of different vendor types.

5.4.1   SS7 network basic design issues

The SS7 data network of a cellular system does more than just carry messages for performing mobile call delivery. This network also provides the messaging for interswitch handoffs, mobile validations, and feature updates to take place. A cell site will go off the air from time to time, but problems with the SS7 network will tend to have a much larger impact on the service to the total customer base. It is for this reason that this network should be properly designed from the beginning and should be constantly maintained separately from the voice network described above. In many operating companies this aspect of the cellular system gets neglected because it is unfamiliar to many within the engineering department and company as a whole.

The responsibility for designing, operating, maintaining, and expanding the network typically is assigned to the facilities group within the engineering department who may not have the background to successfully perform this type of work. With this in mind, it would be a good idea to have a separate group in the engineering department to focus on this area of the network. This group should be responsible for the SS7 network facilities tracking, SS7 network performance monitoring and troubleshooting, and network expansion. More on this topic in Chap. 10.

Follow industry standard protocols for all network design applications that arise in the design and evolution of the system and avoid any "specialty applications" and "vendor-specific" protocols. These types of application solutions may provide immediate and attractive service and design options but will ultimately lead to incompatibility problems in future designs. For example, the use of the IS41 protocol for interfacing between network nodes is recommended instead of using a vendor-specific protocol. This decision to support the IS41 protocol will greatly assist in future intersystem interface applications where the neighboring network uses a differing switch type. The group within the network engineering department responsible for the data network will be able to provide engineering support for these types of design decisions.

5.4.2 SS7 network basic design parameters

The following is a list of basic design parameters for use in designing, building, monitoring, and expanding an SS7 data network.

1. The maximum recommended utilization for an SS7 data link is 40 percent. The initial network design should be built with enough links to accommodate this specification. In other words do not try to cut costs and build the initial system to operate the SS7 data links at 60 or 80 percent utilization. This would not allow for the redundancy functions of the SS7 protocol to be effective should an outage occur somewhere in the network. Furthermore, an expansion of the net-

work should take place any time an SS7 data link exceeds this 40 percent limit.

2. Follow the ANSI specifications for SS7 network designs. Basically each link set will be made up of at least two individual SS7 data links. Each signaling point (SP) should be connected to the SS7 data network using two link sets, one link set assigned to each of the network signaling transfer points (STPs) for redundancy purposes. See Fig. 5.2 for an example SS7 data network.

3. When designing the network facilities at each of the system MTSOs be sure to incorporate patch points into every data link. These patch points will be used to plug the network protocol analyzer into the individual links for performing more advanced call delivery troubleshooting and system monitoring (see Fig. 5.2).

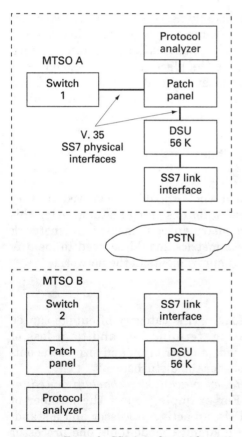

Figure 5.2 Example SS7 interface with monitoring patch points diagram.

4. Separate, as much as possible, the voice interconnection facilities from the SS7 data network facilities. From my experience with working on live cellular systems I have found that by separating these two networks the management of each becomes easier. I tend to protect the data network from the day-to-day operations that take place on the voice network. As noted earlier, a cell site or a T-span may go down for a period of time before it gets fixed with only a small quantity of customers being affected. This type of thinking may be acceptable for voice-type facilities but not in regard to the data network facilities. These demand a higher level of maintenance and problem resolution. For this reason I recommend that these links be placed on separate facilities as much as possible and designated (labeled) as SS7-type data links for the operations department personnel to readily identify.

5.4.3 Cell site data link design issues

The data links of a cell site are usually assigned to one or two channels on the T-span between the base site and the switch. However, the assignment of the cell site data links should still incorporate as much redundancy as possible. When possible, place cell data links on separate switch hardware and transmission facilities. For example, if a cell site has two T-spans interconnecting it to the switch, one data link should be assigned to each T-span.

5.5 Network Reliability and Flexibility

5.5.1 Network reliability

The reliability of the network as a whole should be reviewed at every quarterly engineering design review for designing and improving this aspect of the system. Some of the metrics used to measure the network reliability are given below. These metrics should be used to identify problems and to make reliability improvements in the network.

Network switches

1. *Switch call processing outage.* This category of outage can be broken down into the originating, terminating, and total loss of mobile call delivery service. Provide a description and size of the outage including duration and causes for the disturbance.

2. *Number of switch swaps taken place (use of backup or redundant systems or facilities).* For example, report the number of times a duplex switch (dual CPUs, an active processor and a standby processor) was operating in a simplex mode (only one processor available). Again, provide a description and cause of this switch configuration.

3. *Number of switch billing errors or outages.* The reporting of this metric will require assistance from the revenue assurance group of the accounting and finance department and the information systems department.

SS7 data network

1. *Signaling point (SP) and service switching point (SSP) outages.* This type of outage is defined as the total loss of call delivery service due to an SS7 specific failure mode of an SS7 equipped mobile switch.

2. *Node isolations.* Any network node isolated from the network due to a facilities failure (i.e., cable cut, DAC failure, etc.).

3. *STP failure.* The loss of message-processing capabilities of an STP that affected mobile call delivery service in the network.

4. *SCP failure.* The total loss of message inquiry capabilities by a network SCP.

General network

1. *Network facilities status and failures.* Provide a status of the major network facilities vital to the backup of the system switches and transmission equipment in the event a problem occurs. For instance, are all the generator backup systems in working order? Have they been tested since the last network review? Are the MTSO backup batteries in good working condition? Have they been recently tested along with the system rectifiers in a live cut over test (actually turning off the main power supplies to the switch to allow the battery backup system to provide power)?

Report on the number of failures in any part of these systems along with any possible changes or modifications to these systems that may have recently taken place.

Report on the loading of these systems to evaluate their capacity in light of future network expansions.

Report on the backup systems of the cell sites. Use the criteria mentioned above for reference.

2. *Review and develop plans to improve current network reliability performance.* For example, implement software quality assurance procedures for reducing the number of errors that occur in the switch database and operating software. One such procedure is to standardize the format of the switch database entries. For instance, develop a template of the switch commands required to build a cell site into the switch database and make this available to all network software engineers to reference when performing this type of work. Provide guideline values for each command parameter setting.

Another example might include the increase in the number of back-up tests being conducted in the system. If there are not enough experienced personnel within the company to complete this work, perhaps the equipment vendor of each product could provide this type of service along with a means of reporting their findings.

The reliability of each network is highly dependent upon its size and type of equipment and the environment in which it operates. Therefore, because of these unique issues, it is important to measure and establish your own system level of reliability and set out to improve upon this performance. Owing to the complexity and growth of a wireless system it would be unreasonable to compare its reliability performance to that of an existing and established landline communication system. A wireless operator can get valuable feedback on the reliability of other systems by attending seminars and vendor user forums. These types of meetings also provide valuable data on methods for improving and sustaining the reliability levels of the systems currently in operation.

5.5.2 Network flexibility

The flexibility of a network can be viewed from a number of different standpoints. There is flexibility in the network transmission and equipment designs to allow for various types of interconnections to take place as well as provide the ability for circuit groom and fill functions. There is also flexibility in a network's routing design such that if any part of the network experiences a failure the network should recover to some degree by utilizing its alternate facilities.

To provide flexibility in a network transmission design install DACs at each MTSO to provide DS0, DS1, and DS3 interconnectivity and management. Patch panels should also be included in the MTSO's transmission room along with proper demarcation facilities and documentation.

The ability of a network to recover from a system failure is dependent upon the design of its alternate routes and upon any other backup systems in place across the network. At regularly scheduled times these backup systems need to be checked to assure they are prepared to operate when needed. Development of a routing document should be completed as part of this effort (see Sec. 7.1 for more information).

5.6 Network Management Requirements

The management of a network requires the use of trained personnel and the extensive use of monitoring systems where available. The choice of a network management system should include input from the personnel who will be assigned the responsibility to use the system. Simple as this may sound, this is not always the case. Many times when a network management system (NMS) is purchased it turns out to be the system developed and sold by the same vendor that produces the other network equipment. This may not always be the correct choice. The selection of the NMS should be completed by taking into consideration other outside NMS vendors.

The manual aspects of network management require trained personnel and the development and use of company agreed-upon notification, data collection, and report generation procedures. These are very important and valuable documents developed by the company for use in daily operations and emergency situations.

The design of a network will include many more topics and issues than those presented in this chapter. The concepts shown in these sections are meant to address some of the more important areas to consider when undertaking this task. As you begin to design actual networks the information discussed plus the references listed will provide assistance to assure your design meets the specifications and criteria of the company, the industry standards, and ultimately the service of the expected customers.

References

1. ANSI, 1988, Signalling System Number 7 (SS7), Message Transfer Part (MTP), ANSI T1.111-1988.
2. ANSI, 1988, Signalling System Number 7 (SS7), Operations, Maintenance and Administration Part (OMAP), ANSI T1.116-1990.
3. AT&T Bell Laboratories, *Engineering and Operations in the Bell System,* Murray Hill, N.J., 1984.
4. Bellcore, Bell Communications Research Specification of Signaling System, vol. 1, no. 7, TR-NWT-000246, Issue 2, June 1991.
5. Bellcore, Bell Communications Research Specification of Signaling System, vol. 2, no. 7, TR-NWT-000246, Issue 2, June 1991.
6. Motorola, Inc., *System Description Manual,* 1987.
7. National Engineering Consortium, Inc., *Network Reliability: A Report to the Nation,* Chicago, 1993.

RF System Performance
and Troubleshooting

RF system performance and troubleshooting is one of the most challenging and rewarding aspects of working on a cellular network. There are many aspects to the role of the system performance engineer. One key fact is that no matter how thorough the design work is, it is the system performance engineer who has to make the equipment really work, from an engineering point of view. It is imperative that good engineering practices be employed at system performance troubleshooting. The performance engineer ensures that the lost call and other quality factors are at their best, leading to maximum customer satisfaction and revenue potential.

This chapter discusses system performance and optimization troubleshooting techniques. Topics include how to monitor the network and implement fixes that will be expedient and cost-effective. Pertinent equations are also covered. Direct examples of what has and has not worked are included with explanations. The intention is to impart many of the experiences encountered so problems that have happened in the past may be avoided in the future.

System performance and troubleshooting involve applying a set of critical techniques that have passed the test of time. The first technique is to identify your objective with the effort you are about to partake in and document it. The second technique is to isolate the item you are working on from the other variable parameters involved with the mission statement. The third technique is to identify what aspect of the system you are trying to work on, switch, telco, cell site, mobile, or RF environment. The fourth technique is to establish a battle plan and write down what you want to accomplish, how you will accomplish it, and what the expected results will be. The fifth technique is to communicate what you're about to do and why. The sixth technique is to conduct the work or troubleshooting that is identified in your objective, usually a test plan. The seventh technique is to conduct a postanalysis of your work and then issue a closing document either supporting or refuting your initial conclusions and identifying what are the next actionable items. Summarizing the system performance and troubleshooting methodology:

1. Identify objective.

2. Remove variables.

3. Isolate system components.

4. Test plan.

5. Communicate.

6. Take action.

7. Conduct postanalysis.

6.1   Key Factors

To maximize the RF system performance and expedite the troubleshooting process it is exceptionally important to determine what are the critical system metrics, or key factors, that you need to monitor. You must not only determine what the critical metrics are but also the frequency and level of detail needed. All too often there is either an information overload or underload problem in engineering. Information overload occurs when everyone in the organization is receiving all the reports. Information underload occurs when there are too few reports and they are infrequent. As systems continue to grow in size and complexity the use of statistics for determining the health and well-being of the network becomes more and more crucial. Therefore, when you determine what reports and information you want to see regarding the network it is imperative that a support system is installed. The support system will ensure that the report generation, distribution, and analysis of the data are done on a timely, accurate, and continuous basis.

Most system operators have several key metrics they utilize for monitoring the performance of their networks. The metrics are used for both day-to-day operation and upper-management reports. The particular system metrics utilized by the operator is dependent upon the actual infrastructure manufacturer they are using and the software loads. The metrics that are common to all operators are lost calls, blocking, and access failures. They are very important to monitor and act upon, but how you measure them and ultimately calculate, i.e., the equation, is subject to multiple interpretations. The fundamental problem with having multiple methods for measuring a network is that there is no standard to use and follow.

Numerous metrics need to be monitored, tracked, reported, and ultimately acted upon in a wireless network. The choice of which metrics to use, their frequency of reporting, who gets the reports, and the actual information content largely determines the degree of success an operator has in maintaining and improving the existing system quality. With the proper use of metrics a service provider can be proactive with respect to system performance issues. But when a service-affecting problem occurs in the network it is better for the engineering and operations department to already be aware of the problem and have a solution that they are implementing.

This brings up the issue of what metrics you want to monitor, the frequency with which you look at them, and who receives the information. The key metrics that I have found to be most effective, regardless of the infrastructure vendor or the software load currently being used by the network, are listed below.

1. Lost calls

2. Blocking

3. Access failures

4. Customer complaints

5. Usage/RF loss or usage/number of lost calls

6. Handoff failures

7. RF call-completion ratio

8. Radios out of service

9. Cell site span outage

10. Technician trouble reports

Focusing on the above key parameters for the RF environment will net the largest benefit to any system operator, regardless of the infrastructure they currently are using. When operating a multiple-vendor system for cellular infrastructure you can cross map the individual metrics reported from one vendor and find a corollary to it with another vendor. The objective with cross mapping the metrics enables everyone to operate on a level playing field regarding system performance for a company, and ultimately the industry as a whole. The individual equations for each of the key metrics listed above are discussed in later sections of this chapter.

The key metrics identified above are relatively useless unless you marry them up with the goals and objectives for the department, division, and company. Specifically knowing that you are operating at an access failure level of 2.1 percent, depending on how you calculate it, does not bode well when your objective is 1.0 percent. The decision to use bouncing busy hour versus busy hour for the evaluation also is an important aspect in measuring the system performance. The use of the system-defined busy hour, however, is far better if evaluating trends and identifying problematic areas of the network.

The metrics that you report on need not only to address what you are monitoring but how they are reported. It is very important to produce a regular summary report for various levels of management to see so they know how the system is operating. You should determine in advance when crafting a metrics report who needs to see the information against who wants to see it. More times than not there are many individuals in an organization who request to see large volumes of data, with valid intentions of acting on them, but in the process are overcome with data input so that they enter the paralysis of analysis phase.

What has been very effective in helping maintain and improve system performance in a network involves the establishment of regular, periodic, action plans. System performance is a continuous vigil and requires constant and vigorous attention. Establishing a quarterly and monthly action plan for improving the network is essential in ensuring its health. In particular every 3 months, once a quarter, you should identify the worst 10 percent of your system following the 10 metrics listed above. The focus should be not only by cell site but also by sector, or face, of a particular cell site trying to cross-correlate problems. The quarterly action plan should be used as the driving force for establishing the monthly plans.

Coupled into long-term action plans are the short-term action plans which help drive the success or failure of the overall mission statement for the company. The key to ensuring that the long- and short-term goals are being maintained is the requirement of periodic reports. They are covered in another chapter; however, they are essential, if conducted properly, for ensuring the company's success.

An example of a type of report that will facilitate focusing on the performance of the system is the general format listed below. The metrics listed should be made available to essential personnel on a daily and weekly basis. It is recommended that the following key pieces of information be included in the reporting structure on a weekly basis.

1. Weekly statistics report for the network and the area of responsibility of the engineer trended over the last 3 months.

2. Current top 5 cells and 10 sectors, worst-performing, in the network and each region using the above-mentioned statistics metrics.

3. Listing of the cells and sectors which were reported on the last weekly report with a brief description of the action taken toward each.

4. List of the cells on the current poor-performance list and a brief description as to the possible cause for the poor performance and the action plan to correct the situation.

5. Number of radio channels in the network and by region indicating (a) total number of channels, (b) total number of radios out of service for frequency conflicts, and (c) total number of radios out of service for maintenance.

6. Status of the technician trouble reports (weekly).

The next step in any process is properly using the information that is presented in a timely and useful fashion. For example, when you identify the worst 10 sectors of a system they may be all physically pointing in the same area of the network, indicating a common problem. By overlaying the other metrics listed above in a similar fashion, patterns will appear which will enable the performance department to focus its limited resources on a given area and net the largest benefit.

Figure 6.1 represents a portion of a weekly system statistics report for a network. The chart has only four cell sites listed on it to make the example presented here clearer. The performance criteria for the system involve a lost call rate of 2 percent, attempt failure of 1 percent, and radio blocking between 1 and 2 percent. The chart by itself is interesting, but performing a simple review of the data indicates that a few sectors potentially need investigation. The chart, however, can be converted to a more visual method which aids in the troubleshooting analysis. For the analysis methods presented here a step-by-step approach is shown. The method that you use can combine several of the steps presented. For clarification, however, they are broken out separately here.

1. Sort the chart by lost call percent, focusing on the worst performers.

2. Sort the chart by raw number of lost calls, focusing on the highest raw number.

The reason for the two sorts for the lost calls pertains to how the metrics are calculated. Sorting by the lost call percent alone might not net the largest system performance improvement. The sorting method needs to also incorporate the raw number of lost calls. The lost call percent number can be misleading if there is little usage on the site or a large volume of traffic on the site. If the site has little usage, one

Date:

Sample Busy Hour System Report

Cell Site	Time	Usage	O&T	LC %	# LC	% AF	# AF	%Block	# Block	Usage/LC
1A	1700	262	238	2.1	5	1.26	3	1.26	3	52.38
1B	1700	393	357	7	25	2.52	9	0.84	3	15.71
1C	1700	183	167	1.8	3	4.20	7	1.20	2	61.11
2A	1700	770	700	1	7	0.43	3	—	0	110.00
2B	1700	147	133	1.5	2	1.50	2	1.50	1	73.33
2C	1700	770	700	2	14	1.29	9	0.14	1	55.00
3A	1700	367	333	1.2	4	0.30	1	2.70	9	91.67
3B	1700	419	381	2.1	8	2.10	8	1.05	4	52.38
3C	1700	3438	3125	4	125	0.45	14	1.38	43	27.50
4A	1700	500	455	11	50	0.44	2	0.22	1	10.00
4B	1700	592	667	1.5	10	1.50	10	0.30	2	59.20
4C	1700	183	167	3	5	3.60	6	0.60	1	36.67

Figure 6.1 Partial sample of weekly statistics report for a cellular system.

lost call can potentially represent a 10 percent lost call rate. If the site has a large amount of traffic, it might be operating within the performance criterion, 1.9 percent, but represents 10 percent of the entire system lost calls for the sample period. The resulting display for the lost call percent and raw number of lost calls is shown in Fig. 6.2. There are several sectors, all pointing to a general location using the data from Fig. 6.1.

The next step is to produce a similar chart for the radio blocking statistics. They need to be sorted by percent radio blocking and also raw number of radio blocks. The rationale behind these two sorting methods is exactly the same as that used for the lost call method.

3. Sort the chart by radio blocks percent, focusing on the worst performers.

4. Sort the chart by raw number of radio blocks, focusing on the highest raw number.

The resulting display for the radio blocking percent and raw radio blocking numbers is shown in Fig. 6.3. The information displayed in Fig. 6.3 does not indicate any system level problems.

The next step is to produce a similar chart for the attempt failure statistics. The attempt failure statistics need to be sorted by percent attempt failure and also raw number of attempt failures. The rationale behind these two sorting methods is exactly the same as that used for the lost call method.

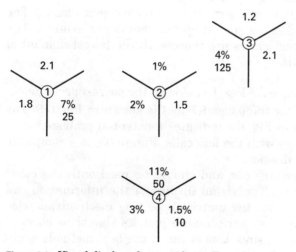

Figure 6.2 Visual display of percent lost calls and number of lost calls by sector.

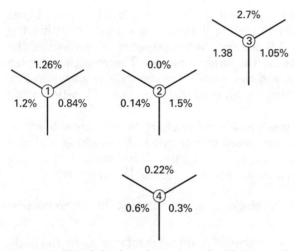

Figure 6.3 Visual display of percent radio blocking and number of radios blocked by sector.

5. Sort the chart by attempt failure percent, focusing on the worst performers.

6. Sort the chart by raw number of attempt failures, focusing on the highest raw number.

The resulting display for the attempt failure percent and raw attempt failure numbers is shown in Fig. 6.4. The information displayed in Fig. 6.4 does not indicate any system level problems.

The next step is to produce a similar chart for usage/RF losses. The usage/RF losses need to be sorted by the worst performers. The usage/RF loss worst performers are those with the lowest amount of usage between lost calls.

7. Sort the chart by usage/RF loss, focusing on the poorest performers.

The resulting display for the usage/RF loss is shown in Fig. 6.5. The information displayed in Fig. 6.5 indicates a potential problem focusing on the same area as with the lost calls. Figure 6.6 is a composite view of the metrics evaluation.

Obviously this procedure can and should be used with the other metrics for the network. The visual display of the information can and should be coupled into the metrics reporting mechanisms used for the network. All the key performance metrics should be checked for any correlation issues, since this is one of the best methods available to identify any performance-related trends.

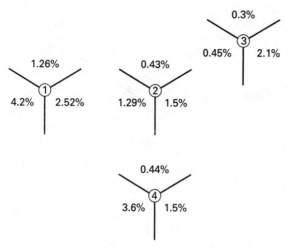

Figure 6.4 Visual display of percent attempt failure and number of attempt failures by sector.

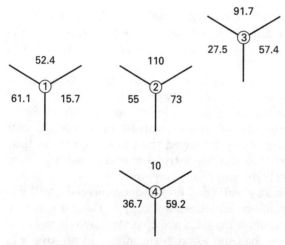

Figure 6.5 Visual display of usage/RF loss by sector.

The reports getting to upper management must be able to tell a story that is both factual and brief. The report needs to allow them to know that the system is running at a particular level but also that you are in control and it does not require their intervention. Many times a senior-level manager who has a technical background, when given too much data, is compelled to generate many questions and inadvertently misdirects the limited resources. The simple rule for dissemination of reports is to minimize the information flow to only

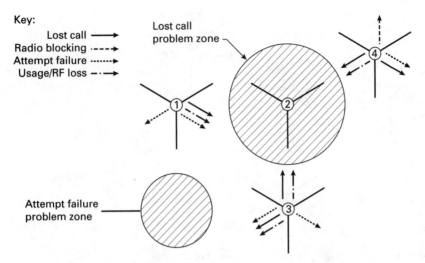

Figure 6.6 Visual display of compilation.

those people who really need to know the material. This is not meant to keep other people in the dark but only to help ensure that the group which needs to continue the focus on improving the network remains doing just that.

However, it is very important to let members of the technical staff know what the current health of the network really is on a regular basis. One very effective method that has been used in various forms is to have the key metrics displayed on a wall for all to see the network's performance. The metrics displayed should be uniform in time scale, and it is recommended that they trend over at least a year so everyone can see how well you are doing over time.

This wall chart, or wailing wall (Fig. 6.7), if done correctly, will foster competition between fellow engineers working on the performance of the network. For example, if the data are displayed by region of the network the engineer who has the worst performance in an area will feel compelled, through peer pressure, not to be on the bottom of the heap for the next reporting period. If this is handled right, the efforts of the various engineers will ensure that the overall performance of the network continues to improve. Remember the old phrase: more flies can be caught with honey than with vinegar.

The other item which needs to be addressed is the need to give regular presentations to the upper management of the technical arena. This is not a classic empowerment method but is meant to stress to the upper management that engineering is performing a good job, hopefully. This will arm upper management with key information so

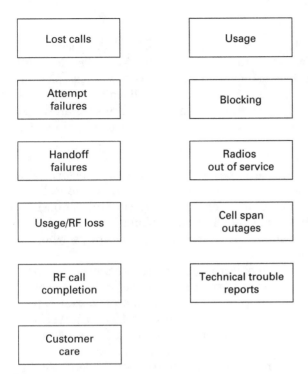

Figure 6.7 Wailing wall.

when they are confronted with irate customers who complain about the lack of good system performance they have some personal knowledge of what is happening.

Reporting to upper management should take place every 6 months and focus on what you have done and will do over the next 6 months. The critical issue here is that if you tell your superiors you are going to do something, make sure you actually do it and report on the next presentation what you did, when, and the results. It is also important that the meeting not take more than 1 hour, so based on the size of the department, i.e., number of presenting engineers, with the time frame well rehearsed and time minimized, focus on three to five items which you can talk about quickly and have the answers for. It is imperative that when you bring up a problem you have a solution that goes hand in hand. The key issue, though, in presenting to upper management is to be truthful and never offer information or proposed solutions to problems that exist when you do not have direct knowledge or control of all the issues.

6.2 Lost Calls

This metric is probably one of the most discussed and focused on parameters in the wireless industry, short of blocking. The lost call metric is personalized because everyone who has a wireless piece of subscriber equipment has experienced a lost call and is personally aware of the frustration and aggravation it causes. Unfortunately lost calls are a fact of life in the wireless industry and will remain so for the foreseeable future.

For many system operators a key component to engineers and upper management incentive is based on the lost call performance numbers. Therefore, how you set the goals and calculate the lost call numbers for the network becomes exceptionally crucial. While it is important to establish a consistent lost call calculation and value to strive for system performance, it is the paying customers who are having their calls dropped.

When focusing on a system objective setting the goal on a raw percentage has merit since it is easily traceable and definable. However, when you equate it to the raw volume of traffic on a network a 2 percent lost call rate for a network can mean over 10,000 physically dropped calls in any given system busy hour for a large network. Therefore, if you had said I have 10,000 lost calls in the busy hour versus 2 percent, the system performance metrics would reflect a different picture but with the same quality impact.

The percentage lost call rate is a valid figure of merit for a system that is growing and has coverage gaps of significant size. The lost call percentage is an extremely effective tool for helping to pinpoint problems and monitor the overall health of any network. However, using the raw lost call percentage number for a large, growing network may in fact be counterproductive, encouraging more actual problems to occur in raw volume than the number represented in percentage indicates.

For example, at a 40 percent system growth rate the 10,000 lost calls at the 2 percent lost call design rate would possibly increase to 14,000 lost calls the next year. Assuming all the parameters increase with the same rate, the lost call rate would remain at 2 percent but the raw number would increase by 40 percent. Obviously this is not what you would want to take place on your own system. An alternative to using the percentage value would be to utilize an additional metric to help define the quality of the network.

The percentage of lost calls should be set so that every year the percentage of lost calls is decreasing as a function of overall usage and increased subscriber penetration levels. Specifically the lost call rates need to be set so that the ultimate goal is 0 percent lost calls in a network. While I do not believe this is feasible at the present state of technology and capital investment, it is still the proper ultimate goal to set and never accept anything less.

But reality dictates that setting of reasonable and realistic goals needs to be done in such a fashion that the real lost call rates are reduced and at the same time the methods utilized are sustainable. How to set the actual lost call rate is an interesting task, since one lost call is too many. The suggested method is to provide at the end of the third quarter of every year the plan for the lost call rate to be striven for in the next year. The goals should be set so that you have a realistic reduction in the lost call rates for the coming year, trying to factor in where you might be at the end of the current year.

The following are a few examples of how to set the lost call objective for your network. Obviously you and your management should be comfortable with the values set forth. The lost call goal should not be set in the vacuum of an office and then downward-directed. However, the goal should be driven to improve the overall performance of the network, factoring into it the growth rate expected, budget constraints, personnel, and the overall network build program.

One example of lost call goal setting involved the situation where several systems were owned by the same overseeing company, which is becoming more common. The interesting issue is that the lost call rates used for goals were different from system to system. The difference in the lost call rate was not necessarily the goal of say 2 percent, but rather the methodology of calculating the equation. This led to an interesting situation where we were able to pick the method which would present the individual company in the best light to the parent company. Specifically one system was using the lost call method of overall lost calls and another was using the call segment approach. The difference between the two is dramatic and needs to be watched for. We chose the call segment method for reporting to the parent company and used the overall lost call rate for an internal method.

The difference between the two techniques is best represented by a simple example using the same system performance statistics:

$$\text{Originations and terminations (O\&T)} = 10{,}000$$

$$\text{Handoffs (HO)} = 10{,}000$$

$$\text{Lost calls (LC)} = 200$$

Overall lost call: $\quad\quad \text{LC/O\&T} = 2\%$
Call segment: $\quad\quad \text{LC/(O\&T + HO)} = 1\%$

The difference between the two methods is rather obvious using the simple numbers listed for the example. Therefore, when you hear or see a lost call percent value the underlying equation and methodology used need to be known to best understand the relevance of the numbers presented.

The different methods of reporting were never resolved nor was an attempt made to resolve them. However, the overall method calculation is not the cat's meow when analyzing system performance since handoffs are part of the overall situation. Some infrastructure vendors do not have the ability to record originations and terminations on a per sector (face) level.

Whether you use the overall method or the call segment method, it is important to be consistent. The relative health of the network can be determined by either method using simple trending methods through comparing past and present performance.

The reported lost call rates to upper management should be visual. Visual methods are extremely useful for conveying a story quickly. However, the visual method can be extremely deceiving whenever a chart or graph is presented. The information can be misleading if the x- and y-axis scales and legends are not defined.

One example of scaling involved two engineering divisions in the same company. They were using the same equations and time frames to calculate their lost call rates. However, both divisions utilized different y axes to display the data. The division with the poorer lost call rate used a larger y axis, having the lost call rate placed in the middle of the chart. The division with the better lost call rate chose a scale which exemplified the lost call rate. The difference in y scales between both divisions was close to a two-to-one margin.

The most interesting point with this example is that the group using the more granular y axis had a lower lost call rate with the higher system usage. However, the perception of upper management was that the division using the more granular reporting scheme was performing the worst since the line on the chart was higher. Many lessons were learned with this example, and one of them was that perception is very important, not facts.

The time used for conducting any lost call analysis should also be clearly delineated. For example, if you use a bounding busy hour method for determining your lost call rate, no matter which equation type you use, it will be difficult to trend. Therefore, when monitoring the trend of the system it is important to establish a standard time to use from day to day and month to month. The standard method of establishing the time to use is the busy hour of the system between the weekdays, Monday to Friday, excluding holidays. The same hour should be used for each of the days for the entire sampling period.

The system busy hour, barring fraud, is usually between 4 and 6 P.M. in the United States. The establishment of the actual system busy hour can be done through a simple analysis of the system traffic usage broken out by hour and day for several months to establish a comfort level that the time picked is valid. It is important to take a snapshot of what the system busy hour is on a regular basis, monthly, to verify that traffic patterns are not changing.

When setting the lost call goals it is important to set the yearly goal and also quarterly goals, at a minimum interval. The issue with setting overall goals and then interval goals solves two primary purposes. The first purpose is to set short-term goals from which to direct the efforts of the company. The second purpose is to ensure that the overall goal is being met, negating the end-of-the-year surprise of either meeting or not meeting the goal. Figure 6.8 is an example of how to set the lost call goal for the network. The lost call rate at the beginning of the year for this example is 2.2 percent and the desired goal is 2 percent. The chart is divided into quarters so a gradual improvement objective can be set. The following are several examples of setting the lost call goals for the network.

One example of how a lost call rate was set involved knowing what the overall corporate goal for the lost call rate was, in terms of a quality figure of merit. The value at hand for this example was a lost call rate of 1.50 percent for the network. The network's current performance, however, was a 2.1 percent lost call rate. Obviously moving from a 2.1 to a 1.5 percent lost call rate in 1 year while experiencing a 40 percent growth rate was not a realistic goal. The unrealistic nature of going for the final number immediately is that this meant an over 28 percent reduction in the current rate, regardless of the method chosen for calculation. Instead the value chosen was 1.9 percent, which involved a 10 percent improvement as a minimum and an overall stretch of 1.8 percent for the final lost call rate. The final numbers were going to be the last month's lost call rate which was to

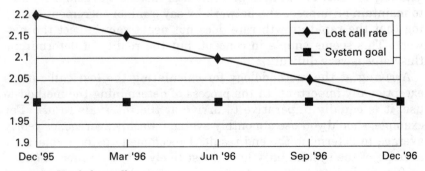

Figure 6.8 Yearly lost calls.

be the total lost call during the busy hours of all the days in the month.

Another example of setting the goals for the department involved upper management picking a value to arrive at. For example, one senior director picked a value for the lost call rate to commit the corporation for the next year's goals and objectives. The goal set was not discussed with any of the management of the engineering department nor with operations to receive their input as to the feasibility of the number being set. The end result is that there was a serious problem with meeting the number, and in fact it was altered during the midpoint of the year. Specifically the altering went as far as redoing the actual goals, which determined the incentives so it could be easily met. Surprisingly no one was held accountable for altering the performance numbers that determine the incentives. The revised number used was the one determined by the engineering department working-level managers.

It is important to set aggressive goals to work for, but it is equally important to involve members of the staff whose job is to ensure that the mission statement is met. In particular what should have happened is that a meeting should have been established to arrive at a value to submit to upper management as to what was realistically expected. The setting of the value without subordinate involvement left the final goal as being the senior director's number and not the number that the department believed in, setting up a fundamental friction point within the organization.

However, when setting the value, there is a midpoint between the examples discussed here. The goal-setting midpoint involves knowing what the lost call rate trend is and determining what percentage improvement is needed for the coming year, factoring in the network subscribers, usage volume, and also cell growth.

For example, setting a goal to reduce the lost call rate by 25 percent from 2 percent, i.e., a goal of 1.5 percent is not realistic if you are growing the network at a 40 percent rate. Instead the goal should be to maintain or improve the network by say a minor increment ensuring that the rapid growth rate does not negatively impact the network. This seems simple in concept, but the reality of determining this value is very difficult.

Arriving at the methodology for calculating the lost call rate is exceptionally important. In the process of determining the method to use, it is equally imperative to arrive at the intervals to use. For example, should you use a monthly average, weekly average, or yearly average to determine the end result? I recommend using a quarterly average of the system busy hour, most likely 4 to 5 P.M. for weekday traffic, excluding weekend traffic. This will eliminate weekly and monthly fluctuations in the lost call rate. It will also compensate for the first few months of the year when the lost call rate is normally lower owing to demographic patterns that seem to always take place. The method will also compensate for a bad cell site software load that is potentially put into the system.

I also recommend using the overall method for deterring the lost call rate of the network, since this is the value the subscriber sees. In a particular call if the subscriber hands off three times and experiences a lost call the lost call rate is only 33 percent using the call segment method. But in the example presented the customer really experiences a lost call rate of 100 percent for that call.

After arriving at the time frame and methodology to use, the actual value will need to be set. The best method for setting the value to arrive at is dependent upon the current level of performance. If your current overall lost call rate is 3 percent during the busy hour, setting the goal of a 33 percent reduction in the lost call rate is reasonable. But if your lost call rate is 2 percent currently, setting a goal of reducing the rate by 33 percent for the coming year means a 1.34 percent level.

This brief discussion has considered only the percentage method of determining the lost call rate to drive for. If you use the raw number of lost calls to have in a network, based on the system growth projection, be careful to look at what the relative performance improvements will be versus the resources available to combat this issue. Specifically if you are currently operating at a lost call rate of 2 percent but the raw number of lost calls during the busy hour is 200, reducing the raw lost calls by 10 percent from the initial number might be a larger percentage when you factor in the actual growth of the network.

At a 40 percent growth rate the O&T level would be 14,000 in the busy hour at the end of the next year, up from 10,000. Reducing the overall lost calls from say 200 to 180 would mean going from a 2 percent lost call rate to a 1.28 percent rate, which is a very aggressive goal. A more realistic goal would be to tie the lost call rate to the system growth level by having the raw lost call rate not exceed an overall percentage number but at the same time not increase and the raw number not increase by more than 50 percent of the system growth rate or stay the same, netting a real improvement. Specifically the goal using this method should be a raw number of 240 as the minimum goal and 220 as the stretch goal to work for.

Whatever the value picked for the benchmark for determining the system health of the network, it is imperative that all the groups involved with achieving the objective help set the goals. However, you need to watch out for the paralysis of a committee deciding the actual value. The end result is that there is no real magical solution to setting the value for lost calls. You must know what your objective is before trying to set the value, i.e., step 1 in the system performance and troubleshooting process.

Once you set your goals to achieve the lost call rate for the system, the next issue you must face is how to identify the poor performers in the network. Many techniques are used, all of which have a certain level of success in trying to reduce the lost call rates. One technique is to use the same report that is used for reporting the health of the network to your upper management and determining what sites are the poorest performers from this set of data. There are two fundamental ways to focus on problems when rating lost call numbers. The first method is to sort the list by poorest performance through sorting by the percentage of lost calls reported on a per sector or cell basis. This list will include the whole system, but you should focus on the top 10 sites at the most for targeting action plans. The second method is to focus on the raw number of lost calls. Both techniques should be used when determining which cell sites to focus attention on first.

For example, using the first percentage method, the poor performers will be identified regardless of the traffic load. The percentage method will help identify if there is a fundamental problem with the site. However, if the cell site has virtually no usage, say 10 calls, and has one lost call the percentage calculation is 10 percent, indicating that there is a serious problem at this site. However, a site operating at a 1.9 percent lost call rate may be contributing 10 percent of the overall lost call rate to the network but since it is such a high-volume cell it shows a lower overall percentage issue by itself. Obviously the focus of attention should be on the site contributing the largest volume of lost calls in this case, not the cell which has the highest percentage of lost calls.

The primary point with the above two examples is that you must think in several dimensions when targeting poorly performing cell sites. The individual lost call rates should be looked at plus the overall impact to the network as a whole when focusing on where to deploy resources.

Another successful technique used in lost call troubleshooting is through utilizing another parameter. The parameter that has found numerous successes in system troubleshooting is usage per number of lost calls on a per sector and per cell basis. This parameter is exceptionally useful for identifying the worst performers in a network, regardless of vendor or software loads used. It will also give you a figure of merit for deterring the level of problems experienced at a site. For example, if you have 5 lost calls with 50 usage minutes, this equates to 1 lost call every 10 minutes. When troubleshooting a system the interval between the lost calls themselves is very important, since the shorter the interval the more problematic the problem is and the higher the probability of finding the root cause in a shorter period of time.

When looking at these three methods of lost call analysis, it is imperative that you look for a pattern. The pattern search is best achieved through a three-step method. The first step is to sort the worst performer by lost call percent. The second step involves next generating a sort by raw lost call numbers. The third step in the process is to sort by the usage/RF loss value. You then take each of the worst 10 or 15 sites, put them onto a map or other visual method, and look for a pattern. More often than not several sites focus on a cluster occurring in a given area. As shown in the examples in Figs. 6.9 to 6.13, the identification of the worst performers nets a pattern of an area that is experiencing a problem. The key issue here is the need to focus on a given area, besides the individual sites involved. The root cause of the problems could be as simple as a handoff table adjustment to a frequency plan problem. There are situations, of course, where there is no real solution to really solve the problem. However, methods are always available to minimize the problem at hand.

Sometimes the problem at hand may not be the site producing poor statistics. One example of this occurred when a system retune took place and a series of problems were reported at one site. Looking at Fig. 6.14, site 1 was reporting that it was producing a high volume of lost calls. Analysis of the data at hand indicated that this was the primary culprit in the problem. The drive team dispatched to investigate

Sample Busy Hour System Report						
Date:						
Cell Site	Time	Usage	O&T	LC %	# LC	Usage/LC
1A	1700	262	238	2.1	5	52.38
1B	1700	393	357	7	25	15.71
1C	1700	183	167	1.8	3	61.11
2A	1700	770	700	1	7	110.00
2B	1700	147	133	1.5	2	73.33
2C	1700	770	700	2	14	55.00
3A	1700	367	333	1.2	4	91.67
3B	1700	419	381	2.1	8	52.38
3C	1700	3438	3125	4	125	27.50
4A	1700	500	455	11	50	10.00
4B	1700	592	667	1.5	10	59.20
4C	1700	183	167	3	5	36.67

Figure 6.9 Lost call statistics.

Sample Busy Hour System Report Date:						
Cell Site	Time	Usage	O&T	LC %	# LC	Usage/LC
4A	1700	500	455	11	50	10.00
1B	1700	393	357	7	25	15.71
3C	1700	3438	3125	4	125	27.50
4C	1700	183	167	3	5	36.67
1A	1700	262	238	2.1	5	52.38
3B	1700	419	381	2.1	8	52.38
2C	1700	770	700	2	14	55.00
1C	1700	183	167	1.8	3	61.11
2B	1700	147	133	1.5	2	73.33
4B	1700	592	667	1.5	10	59.20
3A	1700	367	333	1.2	4	91.67
2A	1700	770	700	1	7	110.00

Figure 6.10   Lost calls sorted by lost call percent.

Sample Busy Hour System Report Date:						
Cell Site	Time	Usage	O&T	LC %	# LC	Usage/LC
4C	1700	501	455	11	50	10.01
1B	1700	393	357	7	25	15.71
3C	1700	743	675	2.1	17	43.68
2A	1700	402	366	4.1	15	26.83
4A	1700	184	167	3	15	12.25
2C	1700	770	700	2	14	55.00
4B	1700	592	667	1.5	10	59.20
3B	1700	220	200	4	8	27.50
1A	1700	367	333	2.1	7	52.38
3A	1700	367	333	1.2	4	91.67
1C	1700	183	167	1.8	3	61.11
2B	1700	147	133	1.5	2	73.33

Figure 6.11   Lost calls sorted by number of lost calls.

the situation confirmed that there was a major problem with the problematic site. Analysis of the cell parameter and neighbor lists of the site itself and the surrounding sites indicated no database or cell site parameter problems.

There was one aberration, however, and that was the low usage on one sector of an adjacent cell site number 2. The site technician was contacted and indicated that the problem being experienced was always there. However, a historic plot (Fig. 6.15) of the site showed a dramatic reduction of the usage on that sector over the same period that the problems appeared at the problematic site. A site visit with the site technician showed that the antenna for a receive antenna was physically disconnected at the antenna input on the tower. This

Sample Busy Hour System Report Date:						
Cell Site	Time	Usage	O&T	LC %	# LC	Usage/LC
4C	1700	501	455	11	50	10.01
4A	1700	184	167	3	15	12.25
1B	1700	393	357	7	25	15.71
2A	1700	402	366	4.1	15	26.83
3B	1700	220	200	4	8	27.50
3C	1700	743	675	2.1	17	43.68
1A	1700	367	333	2.1	7	52.38
2C	1700	770	700	2	14	55.00
4B	1700	592	667	1.5	10	59.20
1C	1700	183	167	1.8	3	61.11
2B	1700	147	133	1.5	2	73.33
3A	1700	367	333	1.2	4	91.67

Figure 6.12   Lost calls sorted by usage per lost call.

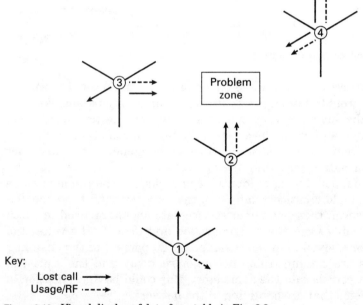

Key:
    Lost call ⟶
    Usage/RF ·---➤

Figure 6.13   Visual display of data from table in Fig. 6.9.

problem was corrected with the aid of an antenna rigging crew immediately dispatched to the site. The problematic site returned to normal operation and traffic levels resumed on the other site.

The example above simply highlights that the problems at an area might not be as directly apparent as just using statistics. Adjacent sites performing poorly, through either low usage or parameter misadjustments, can directly affect the lost call rate.

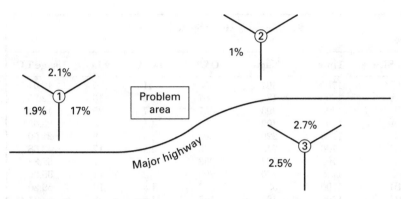

Figure 6.14 Example of problem area.

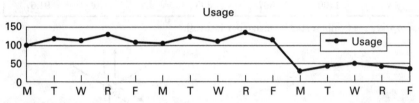

Figure 6.15 Cell site 2, sector 3 usage.

Another successful technique for helping identify areas to focus on for system troubleshooting is to utilize input from customer service. Based on the sophistication that exists between the technical group and customer service the level and volume of information exchanged can range from great to sporadic. It is recommended that customer service information regarding the lost calls reported to them be identified and mapped, trying to identify cluster areas. The problem areas identified should cross-correlate with the data collected from the statistics. However, these customer service data should be used to check the data coming from the metric system and identify the volume of problems for a given area. For example, if 20 percent of the customer complaints are coming from one major roadway and the statistics data do not corroborate the same story, this could be an indication of other problems that are occurring in the network.

The basic items which should be checked for the worst performers in the network are:

1. What changes were made to the network recently in that area?

2. Are there any customer complaints regarding this area?

3. Who are the reusers, cochannel, and adjacent channel for the area, two or three rings out?

4. What is the site's configuration, hardware, antennas, etc.?

5. What are the topology, handoff, and tables for one-way handoffs, either into or out of the site?

6. Are there any unique settings of cell site parameters?

7. Does an access failure problem also exist for the same area?

8. What is the signal level distribution for the mobiles using the site?

9. Is there a maintenance problem with the site?

10. Does one individual radio cause most of the problems?

11. Is a software problem associated with the cell site load?

If you use the above list as a general reminder when initially looking at a poorly performing site, it will expedite your efforts. Most times the problems leap out at you, and of course at times they do not. When the problems are not obvious it is imperative to utilize a checklist. There are many causes of lost calls in a network, and the effort to minimize or eliminate them is an ongoing process. The following are a few more examples of lost call situations that have occurred and continue to occur in one form or another. Several direct and indirect issues can be the cause of lost calls.

One of the more direct causes of lost calls involves mismanagement of the frequency plan where a high level of interference is causing the premature termination of a wireless call. Another direct cause of a lost call can be simply a coverage-related problem in that there is not sufficient signal in the area the mobile is in to ensure continued communication to and from the base station itself. The direct cause of lost calls is probably the easiest to identify, but not necessarily the easiest to fix.

Several indirect issues cause lost calls, and they can be the result of a bad radio channel at a cell site. One bad radio at a cell site can easily be the cause of most of the lost calls at the one cell site. The reason one radio can contribute so much to the lost call problem of a site pertains to its constant availability for use. If the poorly performing radio is causing mobiles to be dropped, it is available more often to receive the next mobile, so it too can be dropped. A poorly performing radio can be caught through normal system diagnostics run every night by operations, but there are no guarantees.

One other example of an indirect lost call situation involves a creation of a handoff hub site. The handoff hub problem is sometimes difficult to detect in a system. Looking at Fig. 6.16, a visual display of the system statistics indicates that there is a handoff problem at the beta face (sector 2) of cell 5. The same chart also indicates that there is a lost call problem with sector 2 of cell 2 in this example. On the

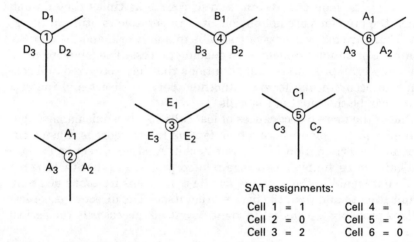

Figure 6.16 Handoff hub.

Figure 6.17 Handoff hub example.

SAT assignments:

Cell 1 = 1   Cell 4 = 1
Cell 2 = 0   Cell 5 = 2
Cell 3 = 2   Cell 6 = 0

surface these two problems would appear to be unrelated, but analysis of the frequency plan (Fig. 6.17) points out an interesting situation. Figure 6.17 indicates that cell sites 6 and 2 share the same channel sets, and the same radio frequencies are used at both cell sites. In addition to the same channel sets being used at these two cells they both are operating with the same supervisor audio tone (SAT).

Figure 6.18 illustrates the root cause of the potential handoff and lost call problems from Fig. 6.16, assuming that cell 5 is in cell 2's handoff list, because it was never removed when cell 1 was added or some terrain issues dictate that this is a valid handoff candidate. A mobile on cell 2 is in need of a handoff to possibly cell 1. Cell 2 issues a handoff request to all the neighboring cell sites in its handoff candi-

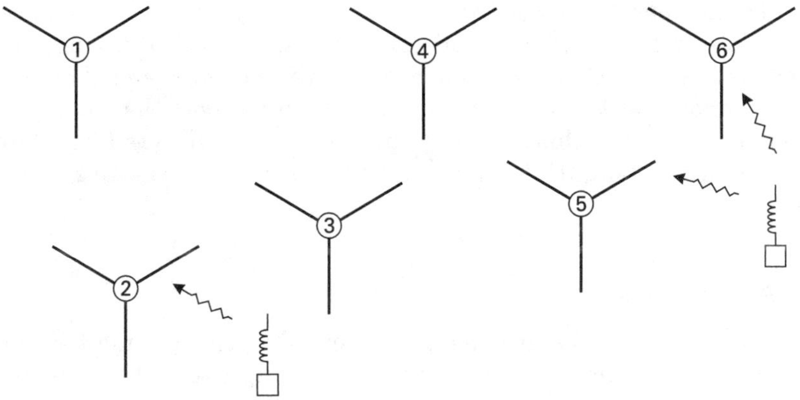

Figure 6.18 Handoff hub.

date table. Cell 5 is also in the candidate table list and scans for the mobile that is on cell 2. However, there is a mobile on cell 6 which cell 5 detects at a strong signal level. Cell 5 reports back, saying it can easily accept the mobile for a handoff, and the system attempts to assign the mobile a channel on sector 2 of cell 5, instead of sector 3 of cell 3. The end result is a lost call and a handoff failure, with the subscriber terminating the call before they wanted to.

There are several corrective actions that can take place for this example. The first step is to remove cell 5 from cell 2's handoff candidate list. The second actionable item is the changing of the supervisory audio tone for either cell 2 or 6.

An example of another indirect type of lost call involves system software installed into a network, related to either cell site or switch module. The ever-changing nature of software in a network necessitates remaining very vigilant regarding ensuring that any software or hardware change made to the network will not negatively impact the subscriber base. It is important to note that not all changes to a network create problems; however, when they do, it can be extremely devastating to both the subscriber and the personnel for the company. For example, upper management noticed on the daily and weekly system statistics reports that the lost call rate for the network was increasing. They of course asked for a reason. The response report stated that a patch was put into one of the switches used in that network on a day in November and that 10 days later it was removed from the network. This action took place before upper management asked for a report.

The lost call rate, after the software patch was removed, changed to the prior rate before the patch was introduced into the network. The good news is that upper management was pleased to see that engineering and operations were on top of the situation. However, the primary problem here is that no reports were issued upward saying that a bad software patch was introduced into the network and that corrective action had taken place. Instead the only report that was generated as a result of this action to upper management was after they questioned the potential problem. The lesson here was to report, report, report.

6.3   Attempt Failure

Attempt failures, also known as access denied levels, are another key metric to monitor and continuously work on improving. The attempt failure level is important to monitor and act on, since it is directly related to revenue. Attempt failures can occur as a result of poor coverage, maintenance problems, parameter settings, or software problems. Regardless of the exact cause for the attempt failure, when you deny a customer access on the network because of its received signal level, this is lost revenue.

The value and methodology used for setting and troubleshooting are very important for determining and improving the health of a network. Most vendors have a software parameter that can be set for establishing the actual received signal strength, or access threshold, level for denying a subscriber service on the network. One constant theme that rises from the ranks of engineering is that the value needs to be set at a level that will ensure a good-quality call, since subscribers would rather experience no service versus marginal service. However, what constitutes marginal service versus good service is very subjective. I argue that the ultimate litmus test for determining what the correct access level is can be derived by monitoring actual system usage and customer care complaints.

The other argument that constantly arises for setting the access parameter is if it is set too low this will increase the lost call rate of the network. Conceptually these two items appear to be directly related. However, in reality the parameter setting and lost calls are not as strongly related as is initially believed. Specifically in a dense urban environment using three different vendors' equipment the lost call rate is not coupled to the access denied level on a one-to-one basis. In fact when access values were set to almost correspond to the noise floor of the system an improvement in the lost call rate still took place at the same time.

It must be cautioned that just setting the threshold parameter to near the noise floor will not necessarily result in reduction in the lost call rate. In fact if the access threshold settings are changed with no other proactive action taken, the net result can easily be an increase in the lost call rate. However, as part of a dedicated program of system performance improvements the attempt failure rate can be successfully reduced at the same time the lost call rate is being reduced.

There are several methods for measuring the attempt failure level in a network. Two of the methods used are identified below.

Method 1:

$$\text{Attempt failures} = \frac{\text{access denied}}{\text{total seizures}} \times 100$$

Method 2:

$$\text{Attempt failures} = \frac{\text{access denied} - \text{directed retries}}{\text{total seizures} - \text{directed retries}} \times 100$$

Using some simple numbers, a comparison of these two methods for calculating attempt failures can be accomplished:

$$\text{Attempt failures} = 1000$$

$$\text{Directed retries} = 500$$

$$\text{Total seizures} = 50,000$$

Method 1:

$$\text{Attempt failures} = \frac{1000}{50,000} \times 100 = 2 \text{ percent}$$

Method 2:

$$\text{Attempt failures} = \frac{1000 - 500}{50,000 - 500} \times 100 = 1.01 \text{ percent}$$

The first method of calculating the access failure rate for the network indicates the true level of problems associated with access issues. However, if you utilize method 2 for your attempt failure calculation the actual level of problems being experienced can easily be misleading.

Another comment about method 2 pertains to the directed retry value reported by the network. Specifically the directed retry will cause another attempt failure at the new target cell site and also add to the total seizure count for the network. The use of directed retry for any reason has to be tightly controlled and monitored to ensure adverse system performance does not take place because it can easily mask system problems. The primary lesson you can derive from this simple example is once again that it is exceptionally important to understand the equation used in the metric calculation.

There are fundamentally two methods of deploying directed retry in a network. The most common method is to employ directed retry at a cell site when it is experiencing blocking. The other method is to redirect a mobile who has not been successful in gaining access to the system as a result of its received signal strength indicator (RSSI) value detected at the target cell site.

For example, using this equation when a subscriber is trying to gain access to the network would peg an access-level attempt on one site which would then get directed to another site and more than likely get denied again. Redirecting a call because of poor signal level is a mistake in system design. Specifically a mobile when gaining initial access onto a system will select the strongest cell, based on forward channel RSSI. Once the mobile determines the strongest cell, assuming a balanced link budget, when a mobile is denied access to one cell due to RSSI at the site, the chance of the situation's improving at the directed retry site is an almost certain failure.

The above example is best shown in a quick series of figures which will help drive the point home. Figure 6.19 shows a subscriber unit preparing to seize a control channel for originating a call or responding to a page. The subscriber unit scans the available control channels and selects the control channel with the strongest RSSI received. The cell site with the strongest RSSI value measured by the subscriber unit is cell A, with a value of −105 dBm. The subscriber unit seizes the reverse control channel (RECC) and attempts to gain access to cell A (Fig. 6.20). Cell A receives the signal sent by the subscriber unit and measures it at a −105 dBm level. Cell site A, however, has an access threshold setting of −100 dBm and denies the subscriber unit access to the system owing to its RSSI (Fig. 6.21). Cell site A instructs the subscriber unit to try to gain access on another cell site through use of a directed retry message sent out on the forward control channel to the subscriber unit.

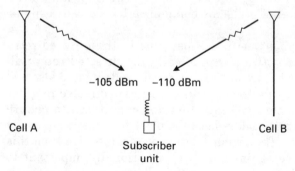

−105 dBm   −110 dBm

Cell A

Subscriber
unit

Cell B

Figure 6.19 Access example.

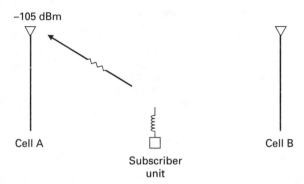

Figure 6.20   Access reverse control channel.

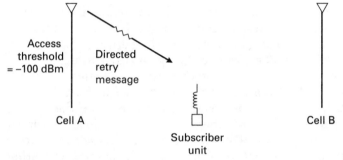

Figure 6.21   Access redirected.

Cell site B now receives a request by the subscriber unit to gain access onto the network. However, the access threshold setting for cell site B is −100 dBm, and the call setup attempt is denied for the subscriber (Fig. 6.22). The net result is that the subscriber was ultimately denied access on both the best possible serving cell for the area and the second best serving cell site. Directed retrying of a subscriber unit based on poor signal level is a parameter that is available for use by several vendors. It is important to determine if you are using this parameter and then understand why you are using it.

Most cellular systems at this time have the ability to deny access to their network by measuring the RSSI value at the intended target cell site. The differences in operating philosophy between sister companies and even competitors with regard to using the access failure parameter are dramatic. For example, one operator was running at a value of less than 1 percent for an access threshold level while the competition was operating at close to 20 percent. Both system operators were proud of the performance of the network and the design levels they were achieving. The interesting point is that at a 20 percent level of access denials this equates to a 1 in 5 chance of being denied

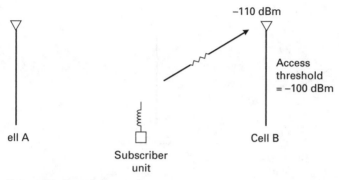

Figure 6.22 Second access.

the ability to place calls on a network. The primary point is that upper management permitted the high volume of attempt failures to take place, largely through not monitoring the network. The 20 percent attempt failure number was not being reported to upper management, since they were using the directed retry method. Engineering, however, was under the misguided belief that their actions were helping the customers and the company in the design, both of which proved to be false. The revenue implications associated with restricting a large volume of call attempts on any network factor directly into the bottom line. However, with the competitor operating at 1 percent, upper management complained to engineering that the parameters were too tight and that any value used for denying access to the network is unacceptable. Many heated discussions took place regarding why you would ever need an access threshold value in the first place. The justification for having a value above the noise floor for this system involved protecting the network from misassigning facilities owing to double originations, or glares.

What value should you adjust the access denied level for in a network? The primary driving force in setting the access threshold value should be revenue. Unless there is a compelling case, clearly identified, the access threshold should be set no higher than 5 dB above the noise floor of the individual cell.

The next logical issue is how you define the system goals and how you monitor your progress and report it to upper management. Like the lost call rate, there are several methods for calculating attempt failures. It is important to keep in mind that any system performance report should reflect what the subscriber experiences.

Two methods of calculating the attempt failure levels in a network were shown earlier. Method 1 is the overall method for determining the access failures on a network. Method 2 is a diluted method for determining access failures onto a network. Whether you use the overall method or the diluted method, it is important to be consistent. The relative health of the network can be determined by either method using simple trending methods comparing past and present performance. However, the diluted method will mask actual problems. If your goal is to improve the overall quality of the network and the revenue potential for the company, the overall method is best.

Several equipment manufacturers provide the ability to report attempt failures on a per sector (face) basis. This is the proper level of granularity for monitoring this important parameter. Obviously the goal ultimately desired is to have no denied access to the network as a result of signal strength levels received. However, until there are no coverage or double access problems in a network an interim value must be used as part of the path for improving the network.

The approach used for setting the value for access failures should be similar to that used for the lost call value method discussed previously. The time frame for the attempt failure rate should be identical for the lost call metrics. The rationale behind this lies in the desire to use these parameters together as a method for qualifying the site's overall performance.

When setting the access failure goals it is important to set yearly and also quarterly goals. Setting overall goals and then interval goals serves two primary purposes. The first in the short term is to direct the efforts of the company through continued incremental improvements. The second is to ensure that the overall goal is being met, negating the end-of-the-year surprise of either meeting or not meeting the goal.

It cannot be overstressed that whatever value is picked as the benchmark to determine the system health it is imperative that all the groups involved with achieving the objective help set the goals. There is no mystical solution to setting the attempt failure value. Before trying to set the value, however, you must determine step 1 in the system performance and troubleshooting process.

Once you have determined the access failure level to use for the system you must face how you identify the poor performers in the network. Many techniques are used, all of which have had a certain level of success in trying to reduce the access failure rates. One technique is to use the same system performance report issued to your upper management. You can then determine what sites are the poorest performers from this set of data. There are two fundamental ways to focus on problems when looking purely at the attempt failure numbers. The first is to sort the list by poorest performance through sorting by the percentage of attempt failures reported on a per sector or cell basis. This list will include the whole system, but you should focus on the top 10 sites at the most for targeting action plans. The second method is to focus on the raw number of access failures and the worst 10 sites in the network. Both techniques should be used when determining the sites to focus attention on first in reducing the attempt failure rate.

For example, using the percentage number the poor performers will be identified, regardless of the traffic load, plus determining if there is a fundamental problem with the site for attempt failures. However, if the site is cosetup co-DCC (co-digital color code) with another site the problem might be not with the site itself but with the actual frequency assignment for the setup channel. With a cosetup co-DCC situation an attempt failure will be recorded by the system for the call being placed, but the subscriber will gain access to the network. The system will report either an attempt failure due to the signal level's being inadequate or a facility assignment failure due to the mobile's not arriving on the target channel.

The above situation is best represented in a series of diagrams. The mobile is near cell site 7 in Fig. 6.23 and originates a call on the network. It scans all the setup channels and determines which cell it is

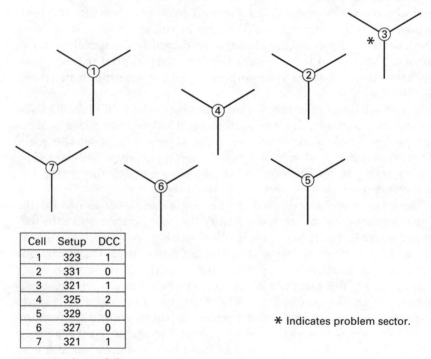

Cell	Setup	DCC
1	323	1
2	331	0
3	321	1
4	325	2
5	329	0
6	327	0
7	321	1

* Indicates problem sector.

Figure 6.23 Access failure.

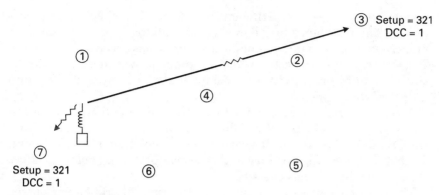

Figure 6.24 Access failure.

receiving the most forward energy on. The mobile selects the strongest channel and responds with a request to access the system on the reverse control channel, 321 for this example. When the subscriber unit responds on the reverse control channel, cell 7 evaluates the mobile for access on the network as part of its call-processing algorithm. However, cell 3 is using the same setup channel and DCC as cell 7. The mobile signal that is sent on the reverse control channel 321 is detected not only by cell 7 but also by cell 3.

In this situation cell 3 will attempt to assign the mobile a voice channel at the same time cell 7 is doing the same. Depending on the system configuration and software a channel could be assigned or an attempt failure would be recorded at cell 3. If cell 3 tries to assign the subscriber a voice channel it will ultimately record a facilities failure for this assignment since the subscriber will have really arrived onto cell 7. In this example the subscriber will have been assigned channels at several potential cell sites at the same time but will successfully arrive at the closer site and be assigned a channel while the problem site identified is reporting problems, but this does not really affect the subscriber. The situation described could be simply resolved by assigning a different DCC value, possibly DCC = 2, for site 3.

An example of another attempt failure problem occurred owing to a serious imbalance in the link budget for a site. The effective radiated power (ERP) for the site was incorrectly specified by engineering for the location, in excess of 200 W ERP when the site should have been operating at a maximum of 50 W ERP. The disparity in signal level was driven by the miscalculation of feedline loss. The operations department set the value for transmit power according to the engineering department's recommended value. The real ERP from the site set up a major disparity in talk-out versus talk-back paths. The additional 6 dB of talk-out power more than the site was designed for resulted in mobiles originating on the site outside of the receive path radius. The imbalance in talk-out versus talk-back paths resulted in numerous attempt failures due to the sign levels received at the site. The situation was corrected when the problem was identified after a site visit to the area with the technician, who noticed the feedline miscalculation.

Access failures can also point to a bad receive antenna where the access failure levels increase right after rain storms and heavy moisture. Since all the sites utilize switching or maximum ratio combining for diversity receive, a bad leg of the receive path can adversely affect the access failure levels of a site.

The examples above point out again that just altering a parameter by itself is not necessarily the solution. Many things can cause an access failure, and it is only through isolation of the variables that the real problem can be uncovered. When trying to minimize the access failure levels in a network, as with other key metrics to monitor, it is imperative that you look for a pattern. The pattern search is best achieved through two methods, both of which you need to use. The first involves sorting the worst performer by access failure percentage number. The next step in the pattern search involves sorting the list by the highest raw access failure. You then take each of the worst 10 or 15 sites and put them onto a map or other visual method and look for a pattern.

An example of pattern searching for access failure problems is shown in Fig. 6.25. Plotting the simple example using the visual method indicates that a problem zone is located between several cell sites. The exact nature of the problem represented in this example might be a coverage issue. The sites that are on the system periphery, if there really is anything like that, might be discarded as being coverage-related. However, in a denser area the identification of a cluster

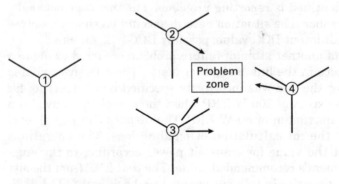

Figure 6.25   Access failure problem zone.

or pattern is very effective for helping troubleshoot the system. The information on the map in Fig. 6.25 should be checked against the lost call rates in the system to help identify any potential patterns. The basic items which should be checked for the worst performers in the network are:

1. Does a lost call rate problem also exist for the same area?
2. Who are the cosetup co-DCC sites?
3. What is the signal level distribution for the originating signals on the site?
4. Are there customer complaints?
5. Is there a maintenance problem with the site?
6. Does one individual radio cause most of the problems?
7. Is a software problem associated with the cell site load?
8. Were most of the problems caused by one mobile or a class of mobiles?

One of the major problems associated with attempt failures is the lack of a dominant server for the area, often referred to as a random origination location. The primary problem with random originations is that there is no one setup, or control channel, that dominates the area where the problem occurs. The lack of a dominant server leads to mobiles originating on distant cells, creating interference problems in both directions, uplink and downlink. The uplink problem occurs since the mobile is in the wrong geographic area and is now in a position to spew interference into a reusing cell site receiver. The downlink problem occurs simply because the frequency set designed for the area is not being used, or potentially not being used. The downlink problem is more pervasive, since the subscriber unit is the victim because of the transmit power reusing cell site.

Figure 6.26 shows a simple RSSI graph with 3 of a possible 21 control channels measured above threshold values. It is evident that not one channel for the primary area in question controls the zone for any given distance. This problem could be reduced through adjustment in the ERP of the control channel for a cell site by 3 dB.

An example of mobiles originating on the wrong cell site follows. They were experiencing interference because they were not being served by the cell that was designed for that coverage area. After a field test it was documented that several of the sectors for the cell in question had voice channels operating significantly higher than the control channel for the site. The primary concern was to ensure that

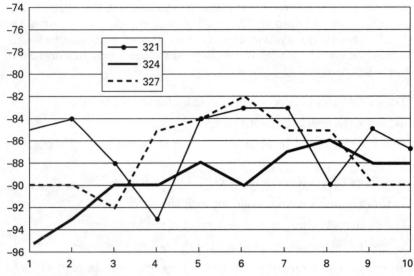

Figure 6.26 No dominant server.

when a mobile tried to gain access to the network it would do so on the correct cell.

In reviewing the little documentation available on the cell and talking with the operating engineers it was pointed out that the control channel was purposely set significantly below the voice channel's ERP to ensure that only good-quality calls are placed on the cell in question. The adjacent cells, however, were not adjusted in a similar fashion. In fact there were varying ERP levels at adjacent cell sites, creating unique coverage patterns for the serving cells in the area (Fig. 6.27).

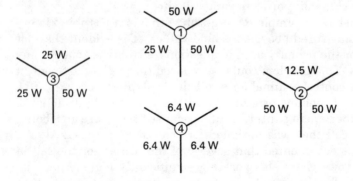

Figure 6.27 ERP setting for each cell site.

The intended coverage area shown in Fig. 6.28 was where the frequency planners expected the site to operate. However, the ERP setting for the site significantly reduced the actual coverage area for the site. The shaded area in Fig. 6.28 shows the area where there is a difference between the origination area for the site and the voice and handoff zones. The disparity in the desired coverage area for the cell site resulted in adjacent cell sites becoming the dominant server in the areas and increasing their effective coverage area outside the designed area where they were to perform. The extension of the adjacent cell sites for setup coverage area resulted in subscribers originating on a cell with a lower than expected C/I ratio.

Analysis of the path imbalance for the site, talk-out versus talkback, did not reflect the level of the problem or the ERP reduction performed at this site. Instead, based on the antennas used and the power level for the site, factoring in the coverage problems north of the site itself, the ERP should have been set so the setup channel coverage matched the voice channel coverage area. What had occurred with this situation was a result of a well-intended action leading to other outcomes. The mismatch of objectives versus results occurred because no posttesting was done to validate or refute the change performed.

One key concept to note on this example is that any reduction or increase in ERP has a dramatic impact on a cell site coverage. While the concept of the relationship between ERP and the cell coverage area seems trivial and obvious, the reality behind it is staggering. For example, reducing a cell's ERP by 3 dB will significantly reduce the

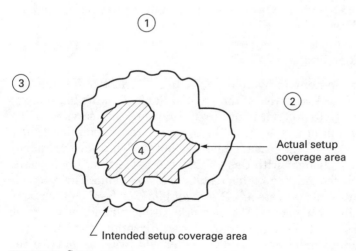

Figure 6.28 Setup coverage.

overall coverage area involved. The relationship is easier to follow if you look at the equation for the area of a cell:

$$\text{Area} = \pi R^2 \tag{6.1}$$

where R = cell radius. This concept eluded two design engineers in that they did not know how to calculate the area of a cell site, assuming a uniform coverage area.

When you adjust the ERP of a cell, the overall area it serves is significantly altered. The usual attention placed on ERP is the forward portion of the link budget, i.e., base to mobile. However, the reverse link needs to also be accounted for since a significant imbalance can easily impact the system's performance in a negative fashion.

Focusing only on the transmit portion of the link budget portion for ERP settings when implementing a 3-dB reduction in ERP for a cell, assuming a 35 dB per decade slope the new cell radius is 82 percent of what it was before. Substituting the new radius into the area equation yields the new cell's effective area as now 67 percent of what it was before the reduction in ERP. Now if you want to increase the ERP by 3 dB using the same 35 dB per decade slope the new radius is 121 percent of what it was before, and this yields an effective area for the cell as being 148 percent of what it was before.

The most fascinating point of this is that when you reduce the coverage of one cell a new dominant server is created, expanding the coverage area of another. Using the single-issue design concept could lead to more attempt failure and lost calls through focusing not on the system but only on an individual cell site. The ERP setting also has an obvious and devastating impact on your capital build program. Another critical point to always remember is that the mobile-to-base path is not changed by an ERP setting.

6.4 Radio Blocking

Another key element in system performance and troubleshooting involves radio blocking levels of the system. Radio blocking has a direct impact on the system performance from a revenue and service quality aspect. Determining just what is the appropriate blocking level for a network has been a subject for many debates. There are primarily three schools of thought with this effort. The first is that any blocking in a network is too much, and this results in lost revenue. The second is that a system should operate at a 2 percent blocking level on the macro level. The third is to operate the network within a band, or range, of blocking and strive to keep the system operating within that band.

The first philosophy has its merit when you want to ensure the most network capacity at a given time. However, this philosophy will lead to overprovisioning of the network in terms of infrastructure equipment, radios, and facilities. The end result is that the inherent operating costs to the network have been substantially increased. One serious downside to this approach, besides the inherent costs associated with the method, is the impact it has on frequency planning. One of the key concepts with frequency planning is to minimize the number of reusers in a given area. This also applies to adjacent channels. If you constantly overprovision the network, the fundamental interference levels will naturally increase since you are using more channels. While there are many techniques for controlling interference, the overprovisioning of channels only makes frequency management more difficult. One additional comment on this approach is that if you desire no blocking on the network this effort involves more than just the RF portion. To ensure no blocking in the network it will be necessary to modify the PSTN blocking level design. For example, if you have a PSTN blocking level of 1 percent it will be very difficult to ensure a no blocking level approach. When establishing a radio blocking level for network operation a complete evaluation of all the components in the call processing chain needs to be factored into the solution.

The second school of thought is the more common method that is employed where a top end number is used. The top end number is a ceiling which is the not to exceed level. The ceiling method is a valid approach when initially deploying a system that is experiencing massive growth and expansion. The ceiling method is also effective when there is very limited labor to monitor and adjust a system's blocking level. The ceiling approach has been successfully used by many operators for designing a network. The ceiling method approach is simple and straightforward. The only real issues with the ceiling level approach are defining the blocking levels, which equations to use, the utilization rate, and the time intervals. The most common blocking level that is used for designing is the 2 percent blocking level for the busy hour. The other particulars associated with the ceiling level approach which need to be focused on will be discussed later.

The third approach to blocking level designs is more relevant to a mature system and requires constant monitoring and adjustments. This third approach, banding, is a very efficient way to utilize a network's resources. The banding method involves setting a top end blocking level a lower end blocking level to operate at. The customary range to operate the banding level approach is to set the upper range at 2 percent and the lower range at 1 percent (Fig. 6.29). Based on the traffic and growth of the network the adjustments needed for the network would seem daunting. However, there is a method that can be and has been used to keep the amount of fluctuations in network pro-

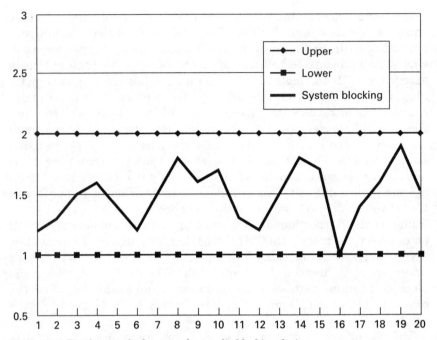

Figure 6.29 Banding method approach to radio blocking design.

visioning requirements to a minimum. Implementing the banding approach will involve a dedicated and focused approach to keeping the system operating within the limits set, however.

The banding approach will also focus on removing channels from cell sites that are not exhibiting high usage. The advantage with removing unneeded radio channels is that it will make available additional channels for reuse, or just simply reduce the aggregate interference levels in the network. Regardless of the method utilized, ceiling or banding, for setting the blocking level of the network, several similar methods of operation need to be adhered to. The items to focus on for keeping system blocking levels within some design guidelines are which equations the utilization rate, time intervals, and logistics management used.

Blocking calculations take on several forms; however, there has been and will be a constant running debate between erlang B and poisson. The equations for erlang and poisson are listed below for reference. Erlang B is the normal industry standard and should be applied for determining the trigger points to act from. A key concept in all traffic engineering is the definition of an erlang. The erlang is a dimensionless unit since the numerator and denominator are of the same unit value and thus cancel out. In the wireless industry an erlang is simply defined as 3600 seconds of usage in a 1-hour time period. Therefore, if you have two radios and each is occupied for 1 hour, 3600 seconds, this represents 2 erlangs of usage.

The erlang B equation is listed below for reference. There are numerous books and technical articles regarding erlang B's statistical nature and accuracy. However, the primary driving element with erlang B is that it estimates, statistically, the probability that all circuits will be busy when you try to make a call. The issue with the erlang B calculation is that it assumes the call, if not assigned a circuit, is lost from the network permanently. Equation (6.2) for erlang B is as follows:

$$\text{Grade of service} = \sum_{n=0}^{n} \frac{E^n}{n!} \tag{6.2}$$

where E = erlang traffic and N = number of trunks (voice channels) in group

The poisson equation is also listed below for reference and should be compared with the erlang B value arrived at for the same number of circuits and grade of service. The fundamental difference between erlang B and poisson is that in poisson you are assuming blocked calls get put into a queue instead of being discarded.

$$\text{Grade of service} = e^{-a} \sum_{n=c}^{\infty} \frac{a^n}{n!} \tag{6.3}$$

where a = erlang traffic
c = required number of servers
n = index of number of arriving calls

Obviously neither erlang B nor poisson represents the real world as it pertains to wireless. Since neither erlang B nor poisson is an exact model for subscriber and traffic patterns in wireless, the continuous debate will remain. However, erlang B is used as the predominant statistical method for traffic calculations and growth predictions. As the foundation for projecting radio blocking values it is the recommended method.

It is very important to know what method of blocking statistics your company is using when projecting radio growth for a network. This has led to some interesting discoveries on how blocking levels are determined. Two primary situations occurred which seriously affected the capital deployment of the networks involved because homemade equations were promoted as erlang B. In one situation the operator was using a traffic table called erlang B. The traffic table was used for defining how many radios needed to be added into a given cell site. A problem was uncovered when putting together a growth study and the desire by engineering to use a spreadsheet for calculating channel growth requirements. Since the traffic table was touted as erlang B, a spreadsheet was structured using Eq. (6.2). The error in the traffic table was uncovered when spreadsheet values were compared to the traffic table provided. The net result was that overprovisioning of radios was being done with the supplied traffic table. After many internal debates the incorrect table was used since it inflated the system requirements to upper management for budget purposes.

The second situation occurred where another operator was using a table, again called erlang B, that was crafted by a traffic engineer. The current traffic engineer when challenged over the table inconsistencies was unaware of what the erlang B equation was and was simply using the traffic table provided. In investigating this issue it was found that there was a deliberate attempt by the engineering manager to overprovision equipment at a cell site to account for a maintenance buffer. The correct erlang B equation was then used by traffic engineering and the number of radios needed for the network was significantly reduced.

The primary point with these two examples is that it is very important to know the baseline equation being used for provisioning the network. An incorrect equation can lead to over- or underprovisioning the network. Specifically you cannot take anything for granted.

The utilization rate for the radio equipment is another key parameter that needs to be focused on and defined. The utilization rate of the radio channels for a cell site sector determines at what point in the process it will be necessary to start provisioning for expansion. When you are at 80 percent utilization rate for a sector, it is time to ensure additional radios are being planned for the site. When you are at 100 percent utilization rate for a cell, based on the derived blocking level used, the turn-around time for restoration is too short. A low utilization rate is indicative of overprovisioning, and a lower limit should be placed on this value to keep the system from being significantly overprovisioned. The utilization rate should be used as a trigger point. The utilization rate picked by you is entirely dependent upon your internal resources, both personnel and financial.

Once you have arrived at a utilization rate to operate from you then perform a trend analysis. It needs to focus on the next period of growth with the purpose of determining the number of radios you will require. The next step in the process is to define the time used for monitoring the network. The same system busy hour that is used for reporting lost calls and attempt failures should be used. However, it is important that a bouncing busy hour traffic report gets generated on a biweekly basis to ensure that there is no fundamental shift in the traffic pattern. The bouncing busy hour report will also help identify fraudulent usage on the network or unique demographics that can help marketing.

This leads to the question of how to go about provisioning the network. The step-by-step generic process is listed below. For this process I assume the banding method using a 1 to 2 percent erlang B blocking level range. In addition the system busy hour and an 80 percent utilization rate are the two other key parameters. The process below can be led by either the manager for network or performance engineering. On a 6-month basis:

1. Determine, based on growth levels, the system needs of 1, 3, 6, and 12 months.

2. Factor into the process expected sites available from the build program, accounting for deloading issues.

3. Establish the number of radios needed to be added or removed from each sector in the network.

4. Modify the quarterly plan that is already issued.

5. Determine the facilities and equipment bays needed to support channel expansions.

6. Inform the frequency planners, performance engineers, equipment engineers, and operations of the requirements.

7. Issue a tracking report showing the status of the sites requiring action. This report should contain as a minimum:

- Radios currently at the site
- Net change in radios
- When exhaustion, 2 percent blocking is expected
- Radios ordered (if needed)
- Facilities ordered (if required)
- Radios secured
- Frequencies issues

- Cell site translations completed
- Facilities secured (if needed)
- Radios installed or removed
- Activation date planned for radios

8. On a quarterly basis conduct a 1-hour meeting to discuss the provisioning requirements and arrange for workers.
9. Perform a biweekly traffic analysis report to validate the quarterly plan and issue the tracking status report at the same time.

The last step in the tracking system process involves the activation date planned for the radios. This particular piece will enable you to preposition the radios in the network without actually activating them. Specifically if the radios are needed for the midpoint of March and are installed in January you can keep the interference levels to a minimum by having them remain out of service.

Blocking problems in a network can occur as a result of a multitude of reasons ranging from normal network growth, a localized traffic jam, changes in traffic patterns due to construction detours, hardware problems, or software problems, to mention a few possibilities. An example of a software problem involving blocking occurred as a result of a system conversion. After 2 months of troubleshooting traffic analysis indicated that the overall system blocking levels were much higher than before the conversion, even after the addition of a significant number of radio channels to the network. The justification for the blocking increase claimed it was a result of more usage on the network because the conversion went so well. However, careful study of the traffic concluded that the increase in blocking did not track the increase in usage (erlangs) for the system for any of the areas in question.

An investigation was then carried out, and through some simple detective work it was found that there was a software problem with the radio assignments. The problem was uncovered through a test where several hundred calls were placed in a designated area and the number of successful calls were recorded. Traffic monitoring during this time was carried out through software and physical inspection and it was noted that several of the calls placed were never assigned a channel due to blocking. The statistics for the system reported the blocking and the subscriber unit experienced the problem.

During this test over half the radios at the site were unoccupied, which conflicted with the data provided for the testing. It was determined that there was a software problem with the particular cell site load causing the blocking problem. The blocking level increase caused by the software problem was affecting both service and capital. The bad load not only aggravated the subscribers but also forced the operator to add additional facilities to the cell sites. The software problem was corrected, but the lesson learned in this situation is that the key parameters used for monitoring the network needed to be tested on a regular basis.

Another pressing issue that also has a negative impact on performance and is usually brought to engineering attention is fraud. The fraud issue for cellular systems at this time is pervasive and has many negative impacts to the network. One key negative is that most operators are forced to build additional capacity into the network functionally to support the fraud carried by the network. One example of how fraudulent use was initially discovered involved a cell site that was experiencing a high volume of blocking. The site blocking levels were largely attributed to the low number of channels assigned to the site. It was believed that with the next wave of system adjustments more channels would be added to the cell and would eliminate the problem.

When the channels were added to the cell site, no blocking relief was monitored. In fact the channels were added on a Friday night and the new channels did not resolve the blocking problem on Saturday. Since the blocking levels persisted, an investigation into the exact nature of the problem began. It was found that the blocking levels were a direct result of fraudulent mobiles operating on the cell site. They appeared to be cloned mobiles based on observations made by the engineering department. The security group for the company was then contacted and a field test was initiated to locate the fraudulent mobiles.

The test equipment used for this investigation was a service monitor, the engineering test vehicle, and a magnetic mount antenna. The channels used for the site were known, and a simple triangulation method was used. Based on the triangulation it was estimated that a minimum of 14 mobiles were operating in the location identified. The blocking levels associated with this site represented over 15 percent of the overall system blocking numbers. This blocking level had a negative impact on the performance of the network and the engineering department.

Several arrests were eventually made, but the blocking problem never was overcome since the fraudulent void was filled as quickly as it was corrected. The site, however, was deemed the worst offender in the network and was removed from the statistics reporting numbers for the network as a result of this report. It was still reported but as a separate line item so that the blocking levels for the site were not reported with the rest of the system.

6.5 Retunes

Any cellular system will at one time require a frequency adjustment, commonly referred to as a retune. Most mature systems experience different levels of retunes on a regular basis. The success of your frequency plan is directly dependent upon your approach to retunes. The level and scope of the interference control in a network need constant attention, especially with a rapid cell and radio expansion program.

Several approaches in dealing with system retunes can be used, based on your configuration and experience level. Retunes can and do take on many shapes and forms. Depending on your own perspective a retune can be viewed as a fundamental design flaw or as part of the ongoing system improvement process. I firmly believe that retunes are part of an ongoing system improvement process. The rationale behind this approach is that there is no real grid, contrary to the popular belief. The primary driving point of the lack of a true grid focuses on real estate acquisition. There is rarely a site selected and built in a network where some level of compromise is achieved with it from the RF point of view alone.

The typical retunes that take place which most people associate with cellular systems involve adjustment to the surrounding sites when a new site is introduced into the network. Other retunes are a result of problems found in the network where there is a cochannel or adjacent channel interference problem. Based on the volume of new sites being introduced into the network the frequency of problems will most likely track with them. With every site added into the network, adjustments can either help or hinder future expansions for the area, either new cells or radio expansions.

Several methods are used for retunes; each has its pros and cons. The main methods used in retuning cell sites are

1. Systemwide flash cuts

2. Cell-by-cell retunes

3. Sectional retunes

The systemwide cuts were probably the favorite at the beginning of cellular development owing to the size of the networks at the time. The usual comment made during a systemwide flash cut is, "This systemwide cut will put the system on a grid and will eliminate the need for more retunes." However, as long as there is system growth the aspects of no more retunes will never materialize.

Two examples of a systemwide retune involved retuning network which were primarily misdesigned from a frequency planning point of view. Both the system operators were managing the frequencies for the network on a channel-by-channel basis. While this process is adequate in the beginning of a system's life, as more and more channels are introduced into a network frequency management cochannel problems become virtually impossible to correct without a retune. Both systems were literally on the opposite ends of the earth and the operators did not know each other. However, the philosophy for the frequency management that the two operators used for both networks was strongly similar. Both operators chose a systemwide retune to end their temporary frequency management dilemma but the actual methods they chose to implement for the system retune varied drastically.

Operator A chose to retune the network over the period of a weekend. The individual base stations involved manually tuned cavities, which of course involved physical visits prior to autotune cavities or linear amplifiers (LACS). The logistics associated with this effort involved relying on a third party to do all the frequency planning and then devise a method for implementing. The third party in the process was more than happy to design a frequency plan but did not implement the recommended design.

The design was implemented through changing the frequencies of every channel in the network through a tape download using a tape method, meaning the network suffered while the retunes took place. The primary point in this case was that the downside to subscribers was not considered and the network was effectively inoperable for a large section during Saturday of that weekend. An additional problem occurred in this effort which was not discovered until 2 months later, and that was that the individual IF setting for the equipment was done incorrectly. The problems of course were later corrected, but the lack of any real postretune testing and nonadherence to customer impact issues left a lot to be desired.

The other system which chose to implement a systemwide retune had different mistakes that took place during their effort. The retune for this system was staged so that one frequency group at a time was impacted. The offending channels were taken out of service along with the channels being retuned. The interesting point here is that in the retune plan the frequency group b was requested to be taken out of service in the network for the retune. During the retune process it was noticed that a major cell on the periphery of the system was taken out of service, because it used the entire b group for channels. However, this individual site was not being retuned and did not require removal from the network. Some fast-acting operations personnel pointed the issue out to the frequency planning engineer, and the site was restored. The point with this effort is that everything that you do not look at in a retune will most likely be a problem.

The method of performing individual sites for retunes is valid for an RSA-type environment. However, for a rural-service-area metropolitan service area utilizing individual site retunes as the primary method for resolving many of the system frequency management problems is not viable. Specifically the time frame and effort needed to retune an area that has say 50 cells will take multiple visits to the same sites as the adjacent sites or subsequent rings are worked on.

Key parameters that need to be looked at for any retune, besides improved service, are time, opportunity cost, and personnel logistics. When putting together a retune the need for completing it in a timely fashion is important. In addition when you are focusing resources on retunes the same resources are not working on other projects for the network. The personnel logistics, which ties into lost opportunities, will be strained and general maintenance will suffer, which will impact the system performance of the network.

The specific method recommended involves sectioning the network into major quadrants. The shape of the quadrants is directly dependent upon the size and scope of the system and personnel available. For example, if your system has 300 cell sites it would be advisable to section the network into three, possibly four sections for retunes. The direction, i.e., clockwise or counterclockwise, is more dependent upon where you think it is more applicable to begin the efforts. In addition it is necessary to limit the scale of the retune itself. It can be contained by predefining the specific items you desire to change for this retune. This design process has a continuous tendency to increase the size and scope of the retune for a variety of reasons. To prevent the size of the retune from increasing to an unmanageable level it is advisable that the manager or director for the engineering department establish a line in the sand after which no changes can take place. Ultimately the line will be compromised, but it will ensure that the size and scope of the retune will be within the logistics available to be successful.

An example of how to set the line is shown in Figs. 6.30 and 6.31. The retune regions are defined as well as the do not exceed line (DNEL), which is different in area from the quadrants set up for the network. The number of cell sites within the DNEL are more than the retune section itself. The actual starting point for the retune sequence and the rotation pattern is a function of several issues. The first is the number of problems, adjacent system interaction, growth plans, and a wild guess.

When setting up a retune it is important to ensure that everyone you need for the retune effort is involved. Failure to ensure all the

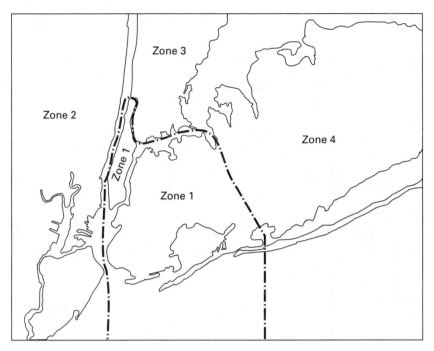

Figure 6.30 Retune zones.

groups are involved with the up-front planning will only complicate life at a later stage. The rationale for involvement lies in the fact that the frequency plan is always a point of contention for every department. Having all the major groups in the process involved will eliminate the finger points and focus on improvement of the network. To ensure that the retune process moves as smoothly as possible, it is recommended that the procedure in Fig. 4.17 be followed.

During the design reviews it is important to ensure that the checklist used for regular retune planning is used for the entire process. The checklist is included below to emphasize that this must be adhered to for success. It is the minimum required to ensure that a proper design review takes place. It is suggested that several reviews take place at different stages of the design process to ensure a smooth integration process. Also the method of procedure (MOP) defined for the retune needs to have internal and external coordination delineated. In addition to all the coordination, the actual people responsible for performing the various tasks need to be clearly identified at the beginning of the process, not in the middle of it.

The following is the checklist for a regional or systemwide retune.

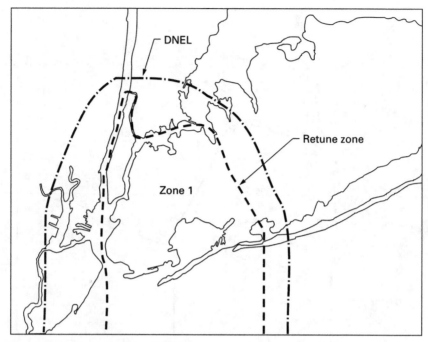

Figure 6.31 Do not exceed line (DNEL).

Voice channel assignments

1. Reason for change
2. Number of radio channels predicted for all sites
3. New sites expected to be added
4. Proposed ERP levels by sector for all sites
5. Coverage prediction plots generated
6. C/I prediction plots generated
7. Cochannel reusers identified by channel and SAT
8. Adjacent channel cell sites identified by channel and SAT
9. SAT assignments checked for cochannel and adjacent channel
10. Link budget balance checked

Control channel assignments

1. Reason for change
2. Coverage prediction plots generated

3. Cochannel C/I plots generated

4. Proposed ERP levels by sector

5. Cocontrol channel reusers identified by channel and DCC

6. Adjacent control channel reusers identified by channel

7. DCC assignments checked for dual originations

8. 333/334 potential conflict checked

Frequency design reviewed by

1. RF design engineer

2. Performance engineer

3. Engineering managers

4. Adjacent markets (if required)

The method recommended is the regional retune process since it takes a systematic approach to frequency management. The systematic approach stresses that there is no perfect frequency plan, and the dynamics of the network in terms of channel and cell site growth necessitate a regular program for correcting the system compromises that are introduced. I have found this method to be very successful in improving the network performance since it ensures that the system compromises introduced into the network over the year can be eliminated or simply improved upon.

The systematic retune process is advantageous and at the same time fraught with many potential downsides. It enables a dedicated group to focus on an area of the network and optimize it to the best of their and the system's ability. Frequency reassignments are not the only factor looked into for a retune. Additional issues are the current and future channel capacity, cell site growth, handoff, and cell site parameters. Basically in a retune you are scrubbing a section of the network involving many aspects besides channel assignments.

The reason you need to look at a multitude of additional parameters for the retune lies in the inherent fact that frequency management is the central cog for cellular engineering. Handoffs, cell site parameter setting, and overall performance of the network are directly impacted by a good or bad frequency plan. Periodically scrubbing a section of the network on a continuous basis will ensure that the compromises made during cell introductions and temporary retunes will be rectified at a predetermined time in the near future. This simple concept will enable the designers to seek more short-term fixes to the multitude of problems they face. Since the systematic retune date for the region is known in advance they can use this future date as a stop point for which to design.

The stop point is important in this process quite simply because how you approach a problem and propose a solution is entirely dependent upon whether it is for 6 months or 5 years. If the solution is for 5 years the time to bring the solution and the volume of unknown variables makes the problem all the more daunting and unmanageable. While it is important to design certain items for 10 to 20 years of useful life, frequency management should be considered short-term.

When conducting regional retunes the scope of the project can easily expand to an unmanageable level. For example, a periodic retune used to help introduce close to two dozen cells occurred. The initial scope of the project involved retuning some 70 cells and introducing the new cells into the network at the same time. During the course of the efforts management lost sight of the actual work being done by the engineers and the overall plan resulted in over 130 cells being retuned and over 100 new channels being added to the existing sites, besides the new cells.

The expansion of the size of the project created a severe strain on both operations and implementation, not to mention the rest of engineering. The strain was caused largely by the frequency planners repeatedly altering their project scope and management not stepping in to stop it. The end result was that operations, as usually is the case, pulled engineering's bacon out of the fire. However, the level of system problems introduced by the strain of the retune took over a month to correct.

Regardless of the method that is used for establishing retunes it is important to always conduct a pre- and posttest. The pretest level for the retune is usually a few drive tests of selected road and a large volume of statistics analysis prior to the switch being thrown. The postretune analysis is one of the most important aspects of the retune process because of the accuracy of the design. It is also important because without it you will miss many opportunities for additional system performance improvements. Most importantly, you will never know if your design efforts were successful and how you can improve upon them. Continuous improvement to the process is the only way you can improve the network on a sustained basis.

One key element that was listed in the retune MOP is the time that the retune will take. It is strongly suggested that when you retune a network the subscriber impact be considered. While this seems rather elementary, the person who is actually paying your salary, the customer, may be forgotten in the heat of the battle to get the task done. It is recommended that all retunes take place in the maintenance window and over a weekend period. This means the fewest negative issues will impact the customers and will allow for maximum time for groups, primarily engineering, to correct any issues that resulted from the retune.

The postretune process that is recommended is listed below.

1. Identification of the key objectives and desired results prior to the retune. For example, the goal may be to facilitate adding five new cells into the network. Identifying key metrics and anticipating their relative change and direction is very important. If you are operating at a 2 percent lost call rate and aiming for a reduction in lost calls by 10 percent to say 1.8 percent with this retune, this might prove to be a bridge too far. It would be difficult if your before and after channel count remained the same, meaning the overall channel reuse in the network would stay about the same. The better objective in this case would be to position the network for future growth without degrading the service levels already there and aiming for improvements in selected zones.

2. Statistics analysis for the 2 weeks prior to the effort, using the same time frames and reference points. One week is the minimum but the more weeks you have in the analysis the better it is to identify a trend. Obviously more than a couple of months of data are not relevant for this effort because traffic, system configuration, software, and seasonable adjustments make comparison very difficult.

3. Full cooperation of operations, implementation, customer service, MIS, and of course engineering for staffing levels. The support you need from each of these groups is critical for your mission to succeed.

4. Staffing during changes. It is important that the crew on duty during the retune document all the problems which occurred during the process and list what they have checked to prevent reinventing the wheel.

5. Postretune statistics analysis. While the drive tests are initially collecting the first tier of data for analysis it is important for the engineers to validate that all the system is configured in a fashion per the design. There has never been a case where problems have not been found during this stage for a multitude of reasons ranging from fat finger mistakes to outright design flaws.

6. Identification of the most problematic areas in the network. Initial statistics analysis is done at this time during or right after the configuration is checked for the network. The problem sites need to be identified by following the key metrics listed before, which are

Lost calls

Attempt failures

Blocks

Channel failures

Usage/RF loss

Customer complaints

Field reports from the drive test team(s)

The initial statistics will focus on an hourly basis owing to the freshness of the retune itself. The data are then checked against the expected problem areas identified before the retune. They are also plotted on a map of the system so patterns can be identified.

This process is repeated every 4 hours for the first 2 days after the retune and then daily for the next 2 weeks. Obviously the degree of detail employed is relevant to the scope of work and the ultimate level of problems encountered.

7. The initial drive test data are then analyzed for the key potential problems. Focus is on the nature of the lost calls and any other problems reported. Problems reported could be dragged calls, interference but no drop, and dropped calls, to mention a few. The problems are then prioritized according to the severity perceived and cross checking with the statistics and anticipated problem areas. An action plan is then put together for each of the problems identified. Sometimes the problem is straightforward, like a handoff table entry missing, or nothing determinable, which requires additional testing as part of phase 3 of the drive testing.

8. Drive test data are then analyzed for the general runs for the rest of the retune area involved, again focusing on any problems that occurred (phase 2).

9. The third phase of testing analysis involves the follow-up tests and postchange corrections needed by the network. If required, additional tests are then performed.

10. Over the next 2 weeks a daily statistics and action report is generated showing the level of changes and activities associated with the effort. This effort is concluded by issuance of a final report.

6.6 Drive Testing

The concept of drive testing a network is usually well understood in terms of its importance and relevance for determining many design issues. Drive testing is used to help define the location of potential cells for the network, integrate new cells into the network, improve the existing network through pre- and postparameter changes, and retune support, to mention but a few. It is always interesting to note that it is usually the drive test team which sets up the test for collecting the data to qualify a potential new site. A well-trained drive test team is exceptionally critical for the success of a network. As most engineers know, any test can be set up to fail or possibly succeed if the right set of conditions are introduced. It is therefore very important to have a strong interaction between various engineering departments and the drive test teams.

Several types of drive tests take place:

1. Presite qualification

2. Site qualification test (SQT)

3. Performance testing

4. Pre- and postchange testing

5. Competition evaluations

6. Postcell turn-on

7. Postretune efforts

Obviously all the items above need to be performed exceptionally well. To ensure that the testing is done well a defined test plan needs to be generated and reviewed with the testers to ensure that they understand what the overall objective really is. Often many alterations to a test plan are left to the test team. If the test team understands what the desired goal and/or results are for the test they have a better chance of ensuring that the alterations and observations they make during the test are beneficial and not detrimental.

One of the key critical elements which needs to be monitored and checked on a periodic basis is the maintenance and calibration of the equipment. Vehicle maintenance needs to be adhered to in order to ensure that the fleet is at key operational readiness at all times. Since major problems in the network requiring full deployment of resources is never a planned event, having the fleet in top operational condition is a high priority. It would be a tragedy if half the fleet, assuming more than two vehicles, were in for some level of repair when a major system problem occurred.

The one area which requires continued attention is the calibration of the field equipment. For the SQT equipment the transmitters used need to be stress tested on a periodic basis to ensure that they are functioning properly for the duration of a test. It is important to ensure that a transmitter used for an SQT that will be operational for say 6 hours maintains the same output power for the duration of the testing. One simple way to check this fact is to ensure that a wattmeter is in line for the test and a before and after test reading is done to ensure that no ERP alteration issues took place during the test.

The antennas and cables used for the SQT need to be validated on a regular basis. A full depot check on all the equipment needs to be performed on a 6-month basis. However, spot checks can be done for each test and recorded in the test log to ensure that these components are operating properly. The cables and antennas used need to be swept on a periodic basis at 3-month intervals to ensure that nothing abnormal has occurred with them. The test should be done on a more regular basis, but with the per test snapshot and the quarterly SQT equipment integrity check the level of confidence with the test equipment functionality should be high.

Calibration of the equipment used for measuring power should follow a yearly calibration schedule as well as the spectrum analyzer, network analyzer, and service monitors utilized. Since so much reliance is placed on the accuracy of the data collected by them it is only logical that the test equipment is checked on a routine basis to ensure its integrity.

Regarding the test vehicle measurement equipment itself this needs to be validated on a more regular basis owing to continued problems which happen to all drive test vehicle equipment. It is strongly recommended that the equipment be tested on a monthly basis using a full calibration test which checks out the functionality of the antennas, receiver sensitivity, RSSI accuracy, adjacent channel selectivity, transmit power, and data deviation, to mention but a few that need to be checked on a regular basis.

The calibration of the test equipment needs to be recorded and stored in a central book that is available for quick inspection by all in the department. The calibration of the field measurement equipment and other SQT pieces needs a comment on the test forms as "in calibration." The calibration records should also include the equipment serial number. In the event that the equipment turns up missing you have a source to track it from; this has proved useful in the past.

Whatever the field measurement equipment used in the drive test vehicle is, it is exceptionally important that you know its accuracy. This involves the adjacent channel selectivity, an important value to know because when you are monitoring say the control channels of the cell sites you obviously are trying to measure an adjacent channel signal level from the dominant server in the area. The ability of the receiver to reject adjacent channel signals is imperative for making rational decisions on problems or potential problems. For example, if you do a single channel plot and notice that it is rather hot in an area near another site the cause might be that the adjacent channel selectivity is not sufficient to isolate the desired from the undesired signal.

The receiver sensitivity is also an important value to know for the equipment since a receiver sensitivity of -102 dBm is not sufficient to measure a signal at say -110 dBm. The difference in the two is an 8-dB signal. In most cases if you are designing for a 17- or 18-dB C/N level the desired serving level might be mistaken for a -85- instead of a -93-dBm, which has a major impact on the design criteria for the area. Many other variables for just the test equipment in terms of how it collects the signals are imperative to understand. Many source documents help determine the sampling intervals required for an accurate RSSI measurement to take place.

The postprocessing of the data is a key area which many people overlook. With the large amount of data collected a significant amount of postprocessing is often done to reduce the amount of data displayed at the end for the engineer to see. Obviously if you average too many points when you are performing an interference analysis a problem area might be masked by the peak interfering signal being averaged with many other bins of data to come out with some nominal level. An example of this would be when driving a large area and going over a bridge or a high-elevation highway where the interference is only a two-block area compared to the 20-mile data-collection run. It is imperative that all postprocessing steps be defined in advance of any data-reduction process so the trade-offs made are understood before they take place.

The SQT role of the drive team is critical in the capital deployment process for any company. As stated before, if the SQT is not done properly a wrong decision can easily be made either to build or not to build a cell site at that location. To ensure the SQT is performed properly it is necessary to establish and follow a test plan. As mentioned in the design guidelines for the SQT, it is imperative that the test plan is made in advance of the site visit by the testing team. The objective for this effort is to ensure that the mounting of the antenna and the other ancillary pieces is done according to the design. The drive routes used for the testing must be sufficiently clear to enable the team to follow the desired direction.

In addition to the SQT is performance testing, which I have labeled pre- and postchange testing. This is a very important aspect of the drive team, and the feedback the group gives the engineers is critical for determining the nature and cause of the problem at hand. Therefore, to ensure maximum output for the drive team the engineer requesting the test must explain to the team exactly what is the test objective. Knowing the objective of the test beforehand will ensure critical feedback is also collected as to any particular issues that arose during the testing. An example of feedback might be when the test team was helping to determine why an area was experiencing a high lost call rate. The fact that part of the major road they drive was almost below grade would have significant influence on the lack of signal for the test area.

For performance drive tests there are functionally two general classifications of tests, internal and external. The two test objectives both focus on trying to identify and help correct a real or perceived problem in the network. Both internal and external performance testing follow the same general testing format. The objective as with all optimization techniques is defining what the objective is and then what you are going to do before taking any action.

Several types of performance tests can be conducted with a drive test. The following is a listing of the major items.

1. Interference testing (cochannel and adjacent channel)

2. Coverage problem identification

3. Customer complaint validation

4. Cell site parameter adjustments

5. Cell site design problems

6. Software change testing

7. Postchange testing

8. Miscellaneous

When scheduling the testing, it is important to identify a priority level for the testing types listed above. The objective of defining a priority level is to have a brief procedure in place when a problem occurs, which it always will, so that the highest-priority level item will be taken care of first. This will prevent the FIFO effect of having drive test scheduling be the only criterion for scheduling tests. Regarding all performance changes made to the network, the need to have a pre- and postchange test conducted can never be overstressed. First the problem is truly identified. The prechange test data should be used as part of the design review process.

Often many good ideas seem great in your head but when you actually put them down on paper and submit them for peer review they may not be as good as initially thought. Once a change is implemented into the network it is equally important to conduct a postchange test, which is meant to verify that the design change worked. The postchange testing is also meant to verify that the changes made to the network do not make any unanticipated problems appear. The postchange test plan needs to also be presented at the time for the design review.

Competitive evaluations are more of an art than a science because the methods used and the benchmarking calculations are primarily nebulous. The variables involved are subjective and define quality. My personal definition of quality is different from your definition. Since there is no specification to truly benchmark against this, the quality figure is left to the determination of an outside consulting firm or the vacuum of a senior management office.

Several attempts to evaluate the current quality of the network and the ranking against the competition have been made. A few companies offer a device which measures the audio levels on both the uplink and downlink paths. The audio level measurements are made in a quantitative method which enables the evaluation to be done the same way time and time again. However, the parameters used in the testing and postprocessing are user-defined, leaving quite a lot open to interpretation. This method will produce the best results since it gets rid of many of the subjective items associated with quality testing.

Another method that is used is to hire people to drive sections of the network and make a series of calls on your system and the competitors' systems. Presumably the calls placed are similar enough in time of day, location, and duration to make a direct comparison. However, the people on the landline or the mobile are left to qualify the quality of the call by listing it as good, poor, or excellent. Obviously the skills and subjectivity of this type of test leave a lot to be desired.

An example of some subjectivity problems occurred once when a quality test came back saying an area of the system was performing exceptionally when engineering and operations knew better. Specifically the area reported to be performing great was a known trouble area which during this time was receiving a lot of attention. The good report was hailed as a success. However, on the next competitive test that was done for the same area the report came back that the area was performing poorly. Upper management was shown the previous test and the current test, which caused much teeth gnashing in engineering and operations. An exceptional amount of effort was then placed on trying to refute the report which previously was hailed as a very good report. The fundamental problem in this case was not the report itself but the fact that when good news is presented, but it is incorrect, it rarely gets challenged. What should have happened here is that the test should have been challenged when it said things were very good knowing that in fact they were not.

The post-turn-on, or activation testing, is exceptionally critical for the insurance that the system is not degraded by the entrance of a new cell site. However, no site has ever been introduced into a network that has not had some level of problem. The argument here is that when no problems are found no one is really looking.

The key to the post-turn-on testing involves two simple principles. The first is that you have to define the test before the site becomes commercial. The second principle is that you need to have the post-turn-on testing done immediately after the new site goes commercial to ensure that problems will be found quickly and rectified. The problems associated with the new site should not be mortal, provided the design guidelines listed in a previous chapter are followed. Instead the issue is the handoff table adjustment, power-level setting, or bias adjustment which needs to take place right after turn-on.

An example of the post-turn-on test request is shown in Fig. 6.32. The post-turn-on test needs to be an integral part of the post-turn-on MOP discussed later in this chapter. The actual form utilized as a trigger point for requesting resources should be well defined. A suggested format to use is proposed, but like all the other forms and procedures presented it is essential that they are crafted to reflect internal organization structures.

The last major area of drive testing involves postretune efforts, which are similar to many of the other tests but done in a tiered approach. Specifically a three-phase approach is recommended, where each phase has a unique mission statement. Phase 1 is the identification and characterization of the most highly probable problem areas for the design, which needs to be validated first. Generally this involves bridges, the major roadways, and areas on the C/I plots

Cell site code: _____

MOP: _____

Requester: _____

Expected drive test start date: _____

Expected drive test start time: _____

Number of test vehicles needed: _____

Estimated drive test duration: _____

Data to be collected: _____

Report any problems to: _____

Special comments: _____

Drive map attached (Y/N)

Figure 6.32 Cell site activation post-turn-on test form.

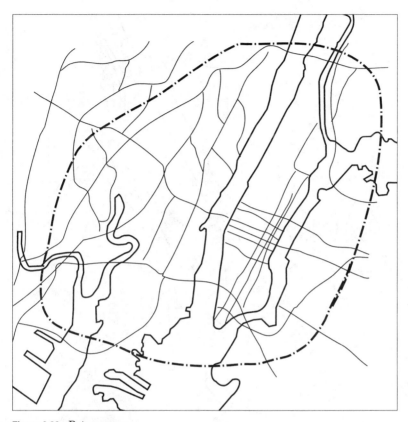

Figure 6.33 Retune zone.

which show anything 19 dB C/I or less, besides gut-level intuition
(Fig. 6.34). Once the potential problem areas are identified a drive
route is designed corresponding to the potential problem areas previously identified.

Phase 2 involves the characterization of the class 1 and class 2
roads in the retune zone. This is meant to try to identify any problems along the major arteries in the network before a full system
load.

Phase 3 involves testing all the areas which showed up as problems
in phase 1 and 2 for either further clarification of the problem or validation that the fixes implemented worked. The issue here is that the
tiered method enables the focusing of resources to resolve the problems in a timely and efficient method.

A key to all the posttesting activities, post-turn-on and postretune,
is that they are conducted immediately after the action takes place.
The rationale behind the immediate time frame is that problems

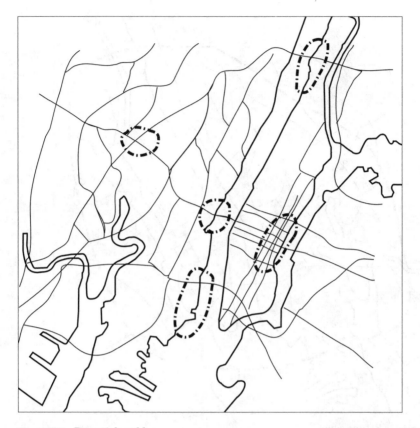

Figure 6.34 Potential problem area.

should be found by the testing team before the subscribers discover them.

6.7 Site Activation

The philosophy of site-turn-on (activation) varies from company to company. Some are driven by engineering and others are driven by financial objectives. Several philosophies are used in the wireless industry. The first philosophy is that once a site is constructed it should be activated into the network. The second primary philosophy is that site depreciation should be minimized or maximized, depending on the accounting method employed by the company. Under the third philosophy the site or sites are not activated until the implementation plan put forth by engineering dictates the timing of the new cells. A fourth site-activation method involves a combination of the second and third philosophies.

Figure 6.35 Phase 1 retune. Drive routes.

The first philosophy of cell site activations, turn it on immediately, has an emotional and upper management appeal. The appeal is that the site is being constructed to resolve some system problem and the sooner it is put into service the sooner the problem the site is designed for will be resolved. While this simple philosophy has its direct merit it also has a few key drawbacks. If done incorrectly it can create more system problems than it was intended to fix. Specifically the drawback involves timing and coordination of the engineering plan to bring the site into the network. If site A requires handoff changes and is activated when the last ATP function is completed the possibility of the handoff changes being implemented at this time is low. One alternative to this is to have the topology changes done in advance of the site activation, but if it is done incorrectly handoff problems and possible lost calls could result. Another situation could occur where there are too many handoff candidates in the topology tables of the sites, complicating the frequency plan for the area.

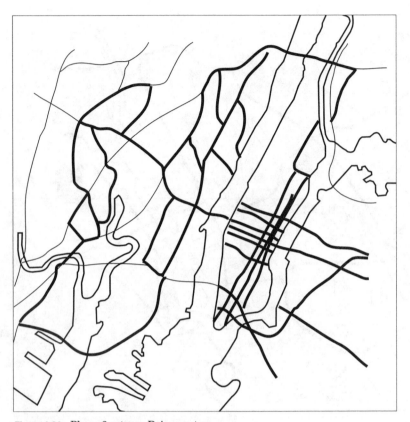

Figure 6.36   Phase 2 retune. Drive routes.

Another major problem with this activation philosophy is the coordination of resources. As in most cases a site needs some level of intra- and intersystem coordination. If the site is activated at a seemingly random time there is no guarantee that all the required coordination has been completed.

The third disadvantage with this method is the queuing of post-turnoff resources for the system troubleshooting phase of a site turn-on. The post-turn-on efforts for their full potential require coordination in terms of timing. If a site is expected to be turned on anywhere within a 3-day window based on implementation problems, it is difficult to ensure that post-turn-on testing will begin right after turn-on. If you want to ensure that post-turn-on testing begins right after turn-on of the cell additional opportunity costs will be associated with this effort, since resources will be significantly mismanaged.

The fourth disadvantage with this effort, the most important aspect, is the customer impact. Turning on when the implementation process is all finished will most likely occur during the day. It is strongly advised that any site or major system action be conducted in the system maintenance window. The simple objective here is to minimize any negative impact the subscriber might experience and try to allow enough turnaround time for the engineering and operations teams to correct any problems before the subscribers experience it.

The second philosophy of activating the sites is to maximize or minimize depreciation costs. This philosophy is largely driven by the financial requirements of the company. Maximizing depreciation costs usually has the operator scrambling to activate as many sites as possible by the end of the fiscal year, usually the calendar year. The objective here is to maximize the potential depreciation expense the company will have in any fiscal year. The opposite philosophy of site activation involves minimizing the depreciation in a given year. This philosophy involves attempts to defer the depreciation of the new cell site into the next fiscal year. What is typically done here is that a site is prepared for activation into the network but will not be turned on until the next fiscal year. The objective is to minimize the amount of expenses reported by the subsidiary to its parent company. Usually the philosophy to maximize depreciation expenses has the activation philosophy of "turn it on now." The minimization of depreciation philosophy usually involves a plan issued by engineering which matches the financial goals of the company.

The third philosophy for site activation is where no new cell site, or system change, is done without a plan being issued and approved by engineering. The objective here is to ensure that the introduction of the new cell into the network has been sufficiently thought out, resources are planned and staged, and all the coordination required has or will be done in concert with the activation. This philosophy is essential to ensure that new cells are introduced into a mature system gracefully.

For this process usually more than one cell site is activated at a time. What is typically done, if logistically possible, is for a series of cell sites in a region to be activated at the same time. This philosophy enables a maximization of post-turn-on resources to focus on the issues at hand and also ensures the minimization of system alterations required. The map in Fig. 6.37 shows how combining several cell sites into a single turn-on for the system will facilitate maximizing resources and minimize the number of changes required to the network.

The largest problem associated with trying to utilize this philosophy is getting upper management approval. The opposition occurs when a site may have to wait several weeks for activation owing to configuration changes needed in the network to ensure its smooth transition. The issue of having a site ready for service and not acti-

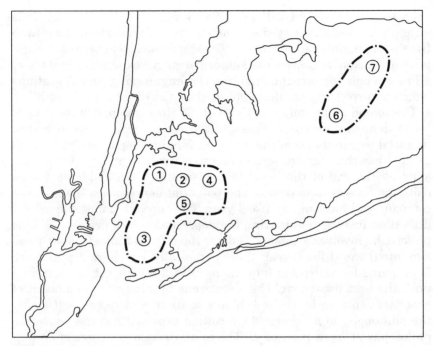

Figure 6.37 Group activation.

vating it immediately is the hardest obstacle to overcome when presenting the case. However, several key advantages that need to be stressed with this philosophy might or might not be apparent:

1. Reduced system problems

    Interference

    Handoffs

    Parameter settings

2. Coordinated efforts between all departments

3. Design reviews of integration plan

4. More fluid intra- and intersystem coordination

5. Pre-turn-on testing

6. Minimized negative customer impact through activation in maintenance window

There are obviously more positive attributes. However, the key issues are reduced system problems, pre-turn-on testing, and minimization of issues that have negative customer impact as a result of following a plan. It is essential that for every cell site brought into a network a plan is generated for its introduction and the plan is then carried out.

The design reviews necessary for a new site activation into the network need to be conducted by several parties. There are several levels of design reviews for this process. The first level involves the RF engineer and the performance engineer discussing the activation plans and reviewing the plan of action. The second level involves having the manager of the RF engineering group sign off on the implementation design with full concurrence of the performance manager. The third level involves reviewing the plan with the director for engineering and operations personnel to ensure that all the pieces are in place and that something has not been left out, like who will do the actual work.

After the design reviews are completed, the MOP for the activations is released. It should be noted that during the design phases the MOP should have been crafted and all of the parties involved informed of their roles. A sample MOP is given in Fig. 6.38 for comparison. Obviously the exact MOP for the situation is different and needs to be individually crafted.

It is essential to always include a backout procedure for cell site activations in the event of a major disaster. The escalation procedure should be defined in the MOP and the decision to go or not to go needs to be at the director level, usually the engineering director or the operations director. After the MOP is released and the design reviews are completed it is essential that the potential new cell be visited by the RF and performance engineers at various stages of the construction period. However, prior to activation it is essential that a pre-turn-on (PTO) take place. The PTO is meant to ensure that the site is configured and installed properly so that when the site is activated into the network the basic integrity of the site is known. The PTO procedure that should be followed is listed later in this chapter.

Internal coordination involving a new site or sites being introduced into the network is essential. The MOP in Fig. 6.38 focuses on voice mail notifications to many groups inside and outside the company. It is essential, however, that the activation of new cells and major system activities be announced to other departments in the company to inform them of the positive efforts being put forth by engineering and operations. The primary groups to ensure some level of notification takes place are

1. Sales

2. Marketing

3. Customer service

Preactivation process

Date

X-X-XX New cell sites to be activated defined
X-X-XX Project leader(s) defined and timetables specified as well as the scope of work
   associated with the project
X-X-XX Phase 1 design review (frequency planning only and RF engineer for site)
X-X-XX Phase 2 design review (all engineering)
X-X-XX Phase 3 design review (operations and engineering)
X-X-XX Phase 4 design review (adjacent markets if applicable)
X-X-XX Frequency assignment and handoff topology sheets given to operations
X-X-XX New cell site integration procedure meeting
X-X-XX Performance evaluation test completed
X-X-XX Executive decision to proceed with new cell site integration
X-X-XX Adjacent markets contacted and informed of decision
X-X-XX Secure post-cell-site activation war room area
X-X-XX Briefing meeting with drive test teams
X-X-XX MIS support group confirms readiness for postprocessing efforts
X-X-XX Customer care and sales notified of impending actions

New cell site activation process (begins X-X-XX at time XXXX)

X-X-XX Operations informs key personnel of new cell site activation results
        Operations personnel conduct brief post-turn-on test to ensure call processing
            is working on every channel and that handoff and handins are occurring
            with the new cell site
        Operations manager notified key personnel of testing results

Post-turn-on process (begins X-X-XX at time XXXX)

        Voice mail message left from engineering indicating status of new cell sites
            (time)
        Begin post-turn-on drive testing phase 1 (time)
        Database check takes place
        Statistics analysis takes place
        Voice mail message left from RF engineering indicating status of postretune
            effort (time)
        Phase 2 of post-turn-on drive testing begins
        Commit decision made with directors for new cell site (time)
        Phase 3 of post-turn-on drive testing begins
        X-X-XX
        Continue drive testing areas affected
        Statistics analysis
        Conduct post-turn-on analysis and corrections where required

X-X-XX
        Post-turn-on closure report produced
        New site files updated and all relevant information about the site transferred
            to performance engineering

Figure 6.38 Method of procedure for new cell site integration.

4. Operations, real estate, and engineering

5. Corporate communications

6. Legal and regulatory

Primarily the entire company needs to be notified of the positive events that take place. One of the most effective methods is through the company's internal voice mail system associated with their extensions. However, not everyone will have an individual extension necessitating the availability of a voice mail account.

To ensure that all the people are notified of the new site activation into the network a series of communications can be accomplished. One method is to issue an electronic mail message to all the employees notifying them of the new sites and any particulars about the intended improvements, if any, to the network. Another method is to slay the trees by issuing a memo to everyone in the company declaring the activation of the sites and the improvements that have arrived.

External coordination for new sites is as essential as internal coordination. Specifically the neighboring systems should and need to know when you are bringing new sites into the network and other major activities. The reason behind this effort is that your actions may have an unintended consequence for them, either positive or negative, which they need to know. Also, by providing your neighboring systems with new site activation information the same level of communication can be reciprocal.

After the site(s) are activated into the network it is essential that post-turn-on testing begin immediately. There has never been a site activated into a network that I am aware of which did not have some type of problem. It is therefore essential that the efforts put forth in this stage of the site activation process receive as much attention as the design phases did.

The key parameters or factors which need to be checked as part of the post-turn-on activities are:

Site configuration checks

Metrics analysis

Drive test analysis

New site performance report

Site configurations from the switch's point of view. The objective here is to check all the cell site parameters for the site as reported by the switch to those intended for the initial design. What you are looking for here is a possible entry mistake or even a design mistake made during the design process. Usually a fat finger mistake is found in this process or an entry is left out. It is imperative that the neighbor cell sites also be checked in this stage of the process.

Metrics analysis.   This part of the post-turn-on activities is meant to help identify and isolate for problem resolution problems reported in the network by the system statistics. This process requires continued attention to detail and an overall view of the network at the same time. The metrics that you should focus on involve the following items:

Lost calls

Blocking

Usage

Access failures

Customer complaints

Usage/RF loss

Handoff failures

RF call completion ratio

Radios out of service

Cell site span outage

Reported field problems called in by technicians or the drive test team

The statistics monitored should be the primary site(s) activated and their neighboring cells. The issue here is to not only look at the sites being brought into service but also to ensure that their introduction did not negatively impact the system. The actual number to use for comparison need to be at least 1 week's prior data for benchmarking, if possible. In addition the numbers used for the new cell should be compared against the design objective to ensure that the site is meeting the stated design objectives.

Drive testing.   The post-turn-on drive test data analysis needs to take place here. This effort usually begins at the specified time after turn-on, usually early in the morning or late at night depending on the activation schedule. The drive tests are broken down into three main categories.

Phase 1 of the driving involves focusing on areas where there is the highest probability of experiencing a system design problem. The identification of these areas can be through use of prior experience, C/I plots, or SWAG.

Phase 2 of the drive testing involves targeting the rest of the areas involved with the site activation activities, usually the remaining class 1 and class 2 roads not already driven.

Phase 3 of the drive testing involves driving areas that were uncovered as problems in phases 1 and 2. This level of testing verifies either that the problem identified previously is still in existence or that the change introduced into the network did its job.

New site performance report. The last stage in the new site activation process is the issuance of the new site performance report. The performance report will have in it all the key design documents associated with the new site. The key design documents associated with this new site should be stored in a central location instead of a collection of people's cubes. The information contained in the report is critical for the next stage of the site's life, which involves ongoing performance and maintenance issues.

To ensure that poor designs do not continue in the network it is essential that the new site meets or exceeds the performance goals set forth for the network. If the site does not meet the requirements, it should remain in the design phase and not the ongoing system operation phase. The concept of not letting the design group pass system problems over to another group is essential if your goal is to improve the network.

The new site performance report needs to include the following items as the minimum set of criteria.

1. Search area request form
2. Site acceptance report
3. New cell site integration MOP
4. Cell site configuration drawing
5. Frequency plan for site
6. Handoff and cell site parameters
7. System performance report indicating the following parameters 1 week after site activation:
   a. Lost calls
   b. Blocking
   c. Access failures
   d. Customer complaints
   e. Usage/RF loss
   f. Handoff failures
   g. RF call completion ratio
   h. Radios out of service

  i. Cell site span outage

  j. Technician trouble reports

 8. FCC site information

 9. FAA clearance analysis

10. EMF power budget

11. Copy of lease

12. Copy of any special planning or zoning board requirements for the site

The new cell site performance report is an essential step in the continued process for system improvements. Only once a site is performing at its predetermined performance criteria should the site make the transition from the design phase to the maintenance phase.

6.8 Site Investigations

Cell site investigations can be equated to a hunting trip. If you are a good tracker and understand the game you are after the chances of success are greatly improved. There are two primary types of site investigations, new sites and existing sites. The new sites referred to here are those which have not begun to process commercial traffic and thus are not activated into the network when visited. The existing cell sites, however, are currently active sites in the network. The existing cell site usually has some particular problem associated with it that now warrants an engineering investigation into the site itself.

6.8.1 New sites

For a new cell site the site investigation is essential to ensure that no new problems are introduced into the network as a result of the physical configuration of the site. The objective of a new site investigation is to validate that the site is built to the design specifications put forth by engineering. Some of the key areas to validate involve

Antenna system orientation

Antenna system integrity

Radio power settings

Cell site parameters

Hardware configuration

Grounding system

The above list of major items can be easily expanded upon and needs to be. The topics listed are primarily engineering issues and do not focus on the radio and cell site commissioning aspects required by operations.

The antenna system orientation is essential to have validated prior to activation. The orientation has a direct impact on how the cell site will interact with the network. For example, if the orientation is off by say 30° then the C/I levels designed for cannot be met. When inspecting the orientation of the antennas it is essential to validate their location and installation versus the AE drawings for the site which were approved by engineering. If there is an obstruction to the antenna system itself, this needs to be corrected quickly.

The orientation of the site can be validated through several methods depending on whether it is a rooftop or tower installation. Ordinarily whoever installs the antennas for the site is required to validate the orientation of the antennas through some visual proof provided the operator. If the installation is a rooftop you should use a 7.5-minute map with some type of optical alignment method, usually a transit. By referencing a point on the 7.5-minute map it is easy to validate the orientation of the antennas with a high degree of accuracy.

When installation is on a tower or monopole the orientation check is a little more difficult. However, the site drawings should reference the orientation of the legs of the tower itself. Once you have the orientation you can verify that the antennas and mounts used are installed in the right locations. However, validating the orientation is more difficult; therefore, you will need to utilize a 7.5-minute map and locate three points at a distance from the tower. Traveling to each of these points, it will be necessary to establish your bearings, and using an optical device (transit), site the antennas for the sector you are in and validate that the bore site for the antenna appears to be correct.

The above addressed physically checking sectored sites, but obviously for an omni cell site orientation is not the issue but rather it is the plumpness of the antenna. It is necessary, where applicable, to also validate the plumpness of every antenna at a site. This can be accomplished through use of a digital level or visual inspection where the application of a level is not practical or safe.

In validating the physical aspects of the antenna system it is necessary to check if downtilt is employed as part of the design and if so what angle is utilized. This can be checked again through use of a set of simple measurements made on the antenna itself. One word of caution: Do not just utilize peg holes for validating the downtilt angle unless the exact angle versus peg hole count is known for that antenna and installation kit. Figure 6.39 is an example of how to calculate the degree of downtilt employed at a site. An additional step in the inspection of the antenna system involves validating that the correct

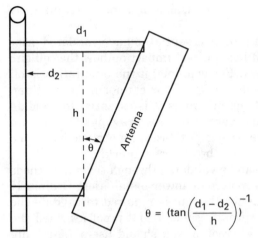

Figure 6.39 Downtilt angle.

$$\theta = (\tan\left(\frac{d_1 - d_2}{h}\right))^{-1}$$

antennas are in fact installed at the site. The simplest method is to verify the make and model number by reading it right off the antenna, where applicable.

One key element to note is physical mounting of the antenna. On several occasions an antenna that has electrical downtilt employed as an omni antenna has been installed upside down since this is a standard installation on a monopole site. The additional interesting aspect of the antenna-inverting situation was that the drain hole was now on the top and the antenna was effectively becoming a rain-level indicator, an interesting sideline but not the intended purpose. Additionally if you are using electrical downtilt and invert the antenna the pattern result is significantly altered for the desired situation, i.e., uptilted.

When checking the antenna system it is important to predetermine the values anticipated prior to actually making the measurements (see Fig. 6.40). For example, if you expect to get a value of 26-dB return loss,

20 dB antenna retune loss

3 dB feedline loss up

<u>3 dB feedline loss back</u>

26 dB system return loss

you would pass the antenna system with a return loss of anywhere from 24 to 26 dB. However, if you got a 14-dB return loss, this would indicate a major system problem since

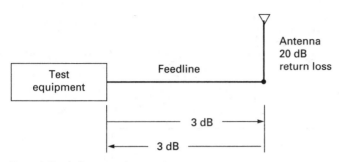

Figure 6.40 Antenna system.

14 dB system return loss

3 dB feedline loss up

<u>3 dB feedline loss back</u>

9 dB antenna return loss

The interesting point here is that most operators will pass an antenna system with a standing-wave ratio (SWR) of 1.5:1, which is a 14-dB return loss. However, when you factor in the cable loss a SWR of 1.5:1 really means that your antenna is experiencing a real SWR of greater than 2.0:1. A SWR of 2.0:1 would have any cellular operator demanding action. Therefore, it is important to remember that the feedline can and will mask the real problem if you are not cognizant of the ramifications.

The last step in the antenna system validation involves validating that the antennas are indeed connected as specified. The real issue is to check if the antenna 1 feedline in the cell site is really connected to the actual antenna 1 for the sector. The simple test involves using a mobile that is keyed on a particular channel and driving to the bore site of every sector, in a directional cell, and making sure that maximum smoke is measured on the antennas assigned for that sector. The key point here is that even the transmit antenna should be checked at this time, provided it is designed to pass energy in the mobile receive band also.

Figure 6.41 depicts a mobile location in the bore site of sector 1 for the cell site. The mobile operator keys the mobile's transmitter and sends energy on a cell site receive frequency. Figure 6.42 is an illustration of what the person at the cell site would be observing with either a spectrum analyzer or a service monitor.

One additional test associated with any cell site involves checking the filtering system used for the site. It is imperative that the actual filter characteristics are checked at this time to prevent additional problems

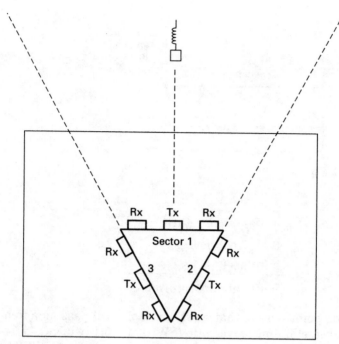

Figure 6.41 Mobile in bore site sector 1.

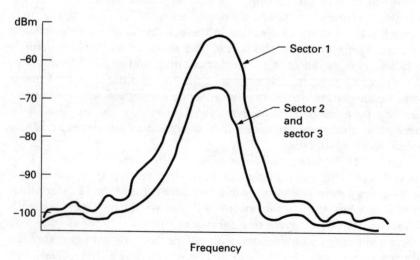

Figure 6.42 dBm display of antenna alignment test.

from being introduced into the network. The most effective method for performing this effort is through a S21 test or rather a through test.

The following is a brief listing of items which need to be checked for a new cell site. Some of the items listed have been talked about; others have not. When reviewing the new cell site checklist it is important to also obtain operations input into the proposal to ensure there is one form and not several. This list should be used as a punch list for correcting identified problems prior to the site going commercial and issues after it has. It is essential that the checklist be used as a minimum and that it be modified and expanded upon based on your own particular system requirements.

New cell site performance checklist

1. Antenna system

   Installation completed

   Installation orientation and mounting verified

   Feedline measurements made and recorded

   Return loss measurements made and acceptable

   Feedlines grounded and waterproofed

2. FAA

   Lighting and marking completed (if required)

   Alarming system installed and operational

3. Receive and transmit filter system

   Bandpass filter performance validated

   Notch filter performance validated (if required)

   Transmit filter performance validated

4. EMF power budget provided

5. Cell site parameters

   Frequency assignments validated

   Handoff topology lists checked

   Cell site parameters checked

   Cell site software load for all devices validated

6. Spectrum check

Sweep of transmit and receive spectrum for potential problems

Colocated transmitter identified

Many of the common problems associated with a new site investigation involve antenna orientation and obstruction issues. It is not uncommon on a rooftop installation to have an antenna obstructed by either a neighboring building or the air-conditioning unit on the roof. Rectification of this situation is to try to relocate the antenna itself or see if possibly using the antenna for another orientation is a better application.

Regarding orientation problems, this situation often involves someone reading a 7.5-minute map incorrectly and orienting the site according to a bad reference point. That is why it is important to have the contractor provide the orientation reference as part of the completion specifications.

6.8.2 Existing cell sites

When trying to improve a system performance it is often necessary to physically visit an existing cell site or location in order to try to determine firsthand the exact nature and cause of the problem for the area. As with all engineering efforts it is essential to have a battle plan laid out prior to performing the mission. It is strongly suggested that the battle plan include possible problems and suggested solutions prior to the field visit itself. With a predefined hypothesis as to the exact nature of the problem you are in a position of either validating or refuting the initial hypothesis.

When conducting site visits to existing cell sites it is recommended that the following checklist of items to include prior to disembarking be done to help expedite the efforts.

1. Review of the statistics for the cell site in question and the surrounding cells

2. Review of the site frequency plan and the surrounding cells

3. Handoff topology review of the site and its neighbors

4. Expected problems to find and possible recommended action items

5. Review of the cell site hardware configuration

6. Review of the maintenance issues for the site over the last month

7. Site access secured for the location

8. Maps of the area and directions to the site secured

9. Test equipment needed for the investigation secured

When defining the site to be investigated it is imperative that a test plan be formulated for the effort. The methods used for defining the area or cell in question are a result of analysis of the key factors that are monitored on a continuous basis. Regardless of the method picked and the nature of the investigation, the following steps need to be taken.

1. Define the objective and specifically what the test or site visit is meant to check for.

2. Identify the area to investigate. Although this is an obvious point the geographic area for investigation is critical to the success or failure of any field test or trip.

3. Ensure that the site picked for the investigation is related to the area defined in step 2.

4. Define prior to the investigation the time you are allocating to investigate the problem, i.e., 2 weeks.

5. Define the physical resources required for the testing, equipment, and personnel.

6. Issue a status report and concluding report on the investigation.

7. Review and update the site-specific books.

Many types of tests can be done for a site, and each requires a different tack to uncover the real problem and recommended fixes.

Interference. One test that could be done is to investigate potential interference at a location. In defining the test it is essential that you identify the potential problems at this location by looking at the following items in addition to the items listed above which are part of the normal site investigation tick list.

1. Current frequency plan

2. Coverage plots for the area (voice and setup)

3. C/I plots for the site and the surrounding area

4. Drive tests

Design issue. Another type of test in an existing cell site could involve investigating potential design problems overlooked in the initial site deployment or later modifications. Some of the items to check for in this part of the investigation are

1. Receive configuration

2. Equipment provisioning aspects for the site

3. IF settings, if applicable

4. Antenna tilt angles looking for overtilt or unbalanced tilting between sectors

5. Antenna elevations, too high or too low

6. Antenna types used

7. Cell site firmware

8. Cell site equipment vintages

9. Setup versus voice channel ERP

10. Cell site parameters

Coverage.   Other site investigation tests involve identifying and qualifying coverage problems. For this type of testing the objective is to determine if there is insufficient coverage in an area that will promote the possibility of interference caused by the lack of coverage. Coverage testing is similar to interference testing in that the basic issues investigated are the same. The recommended use for a coverage investigation involves verifying if the addition of a new cell or changing the antenna system at the site, increasing the ERP, or doing nothing at all is needed.

For cell site parameter alterations and problems a different slant is taken to the site investigation. Testing focuses on evaluation of the handoff window and call processing parameters for the site. A drive test is essential for the pre- and postchange testing, covered previously in this chapter.

Regardless of the type of site investigation taking place some level of documentation needs to detail the findings and recommend actions. The following is an example of a site improvement plan that can be followed or altered based on the particular situation encountered. The situation presented here was an investigation of a site which had its antenna system redesigned. The redesigned antenna system began producing network-related problems, and an investigation into the nature of the problem commenced.

Site improvement plan

1. Generate propagation plots for the site at its current antenna height and at the previous antenna height.

2. Compare the current propagation predictions with the data actually measured.

3. Evaluate data to determine if the coverage presently produced by the site is the desired result for the area.

4. Conduct an S21 test of the antenna system.

5. Conduct a test utilizing the previous antenna system as a comparison.

6. Based on data collected in step 5 determine if further action is required. If it is, determine if the site's previous antenna height should be used employing downtilt or another height should be picked.

7. Implement changes to the antenna system and conduct an S21 test of the new antenna system.

8. Conduct additional field tests to evaluate if the desired results are achieved and make corrections if needed or possible.

9. Evaluate the cell site and its neighbor's statistics performance.

10. Issue a closing report on the engineering activities by (XXX) date.

Figure 6.43 is an example of a site visit report. The cell site investigated was one of the worst five performing cell sites in the network. Several tests conducted with field personnel regarding the site netted marginal results. It was therefore decided to conduct a physical investigation of the site by engineering with operation support.

The key concept to always remember when conducting a site investigation and improvement process is to document what you have done and will do. There is no single cell site that cannot undergo some level of improvement, no matter how small it may seem.

6.9   Intermodulation

Intermodulation situations present themselves to any radio engineer at various stages. The cause of intermodulation and how to remedy the situation has employed many talented engineers and will continue to do so. In order to find and ultimately resolve the intermodulation problem it is important to know the basic concepts of just what intermodulation is. It is the mixing of two or more signals that produce a third or fourth frequency which is undesired. All cell sites produce intermodulation since there is more than one channel at the site. However, the fact that intermodulation products are produced does not mean there is a problem.

Various intermodulation products are shown below for reference. The values used are simple to facilitate the examples: $A = 880$ MHz, $B = 45$ MHz, $C = 931$ MHz, and D is the intermodulation product. The example does not represent all the perturbations possible. Second order:

$$A + B = D \text{ (925 MHz)}$$

Field Test Report

Date: 7-24-95

Subject sites: Cell X

Reason for conducting site visit. The site was chosen to be visited as part of the ongoing process to improve area 3 of the network.

Purpose. The purpose of the site visit was to try to quantify specifically the reasons for the poor performance of sectors 1 and 2.

Site configuration. The site configuration is shown as figure 1, attached. The site consists of four bays of radio equipment and 41 physical radios. There are a total of nine antennas at the site. The site parameters and software load were validated to be correct.

Observations. During the course of the investigation into the site it was noted that there was a serious obstruction to sector 1 and partially for sector 2. Pictures of this situation were taken and are included for reference. It was also found that there was a defective transmit filter for sector 2 of the site, limiting its power output. The filter was acting as a load and therefore did not set off any SWR alarms.

Recommendations. It is recommended that the antenna design for sectors 1 and 2 be reworked to avoid the obstruction. The proposed configuration is shown as figure 2. The defective transmit filter was replaced the day of the site investigation, and no further action is required for this issue.

Engineer: —————————————————

Operations: —————————————————

Figure 6.43   Field test report for a site visit.

$$A - B = D \ (835 \text{ MHz})$$

Third order:

$$A + 2B = D \ (970 \text{ MHz})$$

$$A - 2B = D \ (790 \text{ MHz})$$

$$A + B + C = D \ (1856 \text{ MHz})$$

$$A - B + C = D \ (1766 \text{ MHz})$$

Fifth order:

$$2A - 2B - C = D \ (739 \text{ MHz})$$

The various products that make up the mixing equation determine the order of the potential intermodulation. When troubleshooting an intermodulation problem it is important to prepare for the encounter in advance, if you already know there is a problem. All too often when you conduct an intermodulation study for a cell site numerous potential problems are identified in the report. The key concept to remember is that the intermodulation report you are most likely looking at does not take into account power, modulation, or physical separation between the source and the victim, to mention a few. Therefore, the intermodulation report should be used as a prerequisite for any site visit so you have some potential candidates to investigate.

Intermodulation can also be caused by your own equipment through bad connectors, antennas, or faulty grounding systems. However, the majority of the intermodulation problems encountered were a result of a problem in the antenna system for the site and well within the control of the operator to fix.

Just how do you go about isolating an intermodulation problem is part art and part science. I prefer the scientific approach, since it is consistent and methodical in nature. If you utilize the seven-step approach to troubleshooting listed at the beginning of the chapter you will expedite the time it takes to isolate and resolve the problem. The biggest step is identifying the actual problem, and the rest of the steps will fall in line. Therefore, it is recommended that the following procedure be utilized for intermodulation site investigations.

Previsit work

1. Talk to the cell site technician and have him or her go over the nature of the problem and all the steps taken to correct it.
2. Examine the site-specific records for this location and see if a previous problem was investigated and if any changes were made recently to the site.
3. Determine if there are any colocated transmitters at this facility and conduct an intermodulation report looking for hits in your own band or in another band based on the nature of the problem.
4. Collect statistical information on the site to try to determine any problem patterns.
5. Review maintenance logs for the site.
6. Formulate a hypothesis for the cause of the problem and generate a test plan to follow.
7. Secure the necessary test equipment and operations support for the site investigation.
8. Allocate sufficient time to troubleshoot the problem.

Site work

1. Plan the initial test.

2. Isolate the problem by determining if it is internal or external to the cell site.

3. Verify all connectors are secure and tight.

4. Monitor the spectrum for potential intermodulation products determined from the report.

5. When intermodulation products appear, determine common elements which caused the situation.

Based on the actual problem encountered, the resolution can take on many forms.

If the problem is a stray paging transmitter the recommended source of action is to notify the paging company and request, first, that they resolve the situation immediately. This situation requires a posttest to validate if the change took place and netted the desired result.

If the problem is a bad connector producing wideband interference, the situation is corrected by replacement of the connector itself.

If the problem is a bad duplexer or antenna, again the situation is rectified through replacement of the equipment itself.

If the intermodulation product is caused by the frequency assignment at the cell site, it will be necessary to alter the frequency plan for the site, but first remove the offending channels from service.

If the intermodulation problem is due to receiver overload, the situation can be resolved by placing a notch filter in the receive path if it is caused by a discrete frequency. If the overload is caused by cellular mobiles, using a notch filter card also resolves the situation. Another solution to a cell site overload problem can be resolved by placing an attenuation in the receive path, prior to the first preamp, effectively reducing the sensitivity of the receive system.

Figure 6.44 is a field report for an intermodulation investigation that took place.

6.10 Orientation

The orientation of the sectors in a wireless network directly determines the effectiveness of the frequency management scheme employed by the operator. The consistent orientation of the sectors is critical for getting the maximum C/I ratio available in the frequency assignment. Frequency reuse is maximized through controlling where the potential interference will be. Using different orientations in a network will lead to increases in lost call rates and poor performance due to increased interference. There are numerous technical articles pertaining to the use of orientations and frequency management. The use of standard orientations also facilitates the system performance troubleshooting through the elimination of variables. The orientation of the cell sites is critical not only for frequency management but also for handoffs. Orientation is not a critical factor in a young system because most of the performance problems are directly related to coverage issues. However, coverage issues can be caused by incorrectly orienting antennas at a cell site. The orientation checking procedure is included in Sec. 6.8.

The orientation selection used by the system is usually 0° true north for sector 1 and rotated 120° clockwise for the remaining two sectors for a three-sector cell. Figure 6.45 shows several different system orientations utilized for the same geographic area but different systems. The choice of orientation used is not as important as keeping the orientation the same throughout the network and interfacing it to other networks. The orientations shown in the figure are strikingly different. The configuration shown for system 1 involves two orienta-

---

Field Test Report

Date: X-X-XX

Subject sites: Cell Y

Reason for conducting site visit. The site was chosen to be visited because of an intermodulation problem reported by operations.

Purpose. The purpose of the site visit was to isolate and correct the intermodulation problem at the site.

Site configuration. The site configuration is shown in figure 1, attached. The site consists of five bays of radio equipment and 53 physical radios. There are a total of 12 antennas at the site. The site parameters and software load were validated to be correct.

Observations. The intermodulation test plan conducted 2 weeks prior was repeated, since it was suspected that the original problem identified corrective actions recommended but never done. As reported previously, the intermodulation was wideband in nature, encompassing the entire receive band. A minor amount of investigation identified that the connector used to connect the duplexer to the antenna was still defective. No other testing was done since the problem was immediately identified. The defective connector was removed from the system by replacing the jumper cable it was on.

Recommendations. The defective connector was replaced, and no further action is required.

Engineer: ⎯⎯⎯⎯⎯⎯⎯⎯⎯⎯⎯⎯⎯⎯⎯⎯

Operations: ⎯⎯⎯⎯⎯⎯⎯⎯⎯⎯⎯⎯⎯⎯⎯

Figure 6.44 Field test report for an intermodulation investigation.

tions, one at 0° and the other at 30° true north. The rationale behind the two orientations involves interfacing a three-sector system with six-sector systems to the north and south of the network. The entire system was never oriented at 30° true north because of the road structure in the core of the network. For system 2 the orientation for the network is unique, to say the least. The reason for the orientation picked was never uncovered by later engineering generations since there was no paper trail defining the design decisions made.

Orientation has a major role in handoffs since a poorly oriented antenna will create ping-pong effects or coverage gaps. The ping-pong effects can be corrected through use of parameter adjustments, but the coverage gaps cannot be corrected through use of software.

One particular orientation problem occurred with a six-sector cell site. Drive test data for the area indicated that the call was flipping between several sectors of a cell site, and ultimately dropping. The propagation analysis conducted for the site showed there should be sufficient signal to carry the call with no problem. Field test results showed the control channel coverage matched the predictive evaluation very closely. However, the signal levels for the calls on the site showed a significant problem.

The site's radios and antenna system were checked for power output and SWR with everything coming back clean and green. Engineering then decided to conduct a site visit to the location and brought along a transit. During the course of the field investigation it was noted that two of the sectors failed the one-eyeball test and were checked with the transit. The orientation for both of the sectors was significantly off by close to 20°. Considering the site was a six-sector cell site, this represented an orientation error of some 33 percent. An antenna rigging crew was dispatched to the location and all the antennas were checked for improper orientation and corrected. The net result was that the poor performance of the site was significantly reduced to acceptable levels for the system.

6.11 Downtilting

Downtilting or altering the antenna inclination of a cell site is one of the techniques available to a radio engineer for altering the site coverage. The rationale used to alter the concatenation of the cell site antenna system can be varied. Some reasons for altering involve reducing interference, improving in-building penetration, improving coverage, or limiting a site coverage area. The concatenation of the antenna system can have a major impact on the actual performance of the cell site itself.

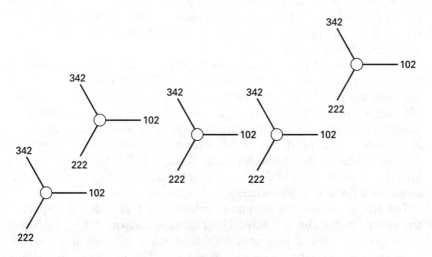

System 1 orientations

System 2 orientations

Figure 6.45   Orientations.

The alteration of the tilt angle for a cell site should be done with extreme care. Altering the antenna system of a cell site can have a major impact, both positive and negative, on the performance of the network. To maximize the benefits of altering the tilt angle for a cell site while minimizing your exposure to problems, it is suggested that you follow a test plan.

Several articles and papers have been written regarding the use of downtilting antenna systems. The primary gist of the reports is the use of tilting the antenna system, usually half the vertical beam width, to achieve a 3-dB reduction signal level at the periphery of the sector coverage. The tilt angles are usually specified, or requested, by the frequency planners for the intention of improving the C/I ratio at another cell site.

I have found that taking the terrain aspects near the cell site can provide large signal attenuation with just a minor tilt angle. The use of terrain to assist in the attenuation of the signal is based on diffraction of the signal. Diffraction on attenuating the signal is a very effective tool when trying to maximize the signal near the cell site and attenuate it near the horizon.

For example, if your objective is to improve the coverage near the cell site but contain the coverage of the cell so it does not create an interference problem to the network, altering of the antenna system tilt angle can be considered as one possible solution, assuming the antenna system employs an antenna with a 5° downtilt. Figure 6.46 shows where the main lobe of the antenna pattern strikes the ground. The main lobe of the antenna system using a 100-ft-high cell site at 5° hits the ground at 1143 ft. The example used for Fig. 6.46, however, assumes a flat earth situation, which is not realistic.

The tilt angle for the antenna system also has a direct impact on the coverage for the cell site. The example shown in Fig. 6.46 indicates that with a cell radius of 2800 ft another cell site might now be required for the area to provide coverage. The coverage loss due to downtilting is often an overlooked aspect when this technique is employed as a solution. It could be reduced or eliminated by utilizing diffraction. The use of diffraction on attenuating the signal for a cell site has a greater impact than merely tilting the antenna by half its vertical beamwidth.

Figure 6.47 is an example of using diffraction to attenuate the signal. The figure shows an obstruction 3000 ft away from the site, bore site for the sector. The desired goal is to the signal level near the cell site but to minimize the negative effects of interference at the reusing cell site, several sites away. The antenna tilt angle needed to achieve over 21 dB of attenuation is less than 2°. The use of terrain to assist in the attenuation of the signal should be exploited in order to maxi-

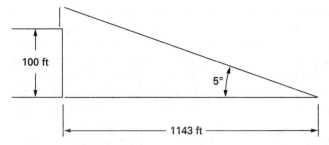

Figure 6.46 Downtilt.

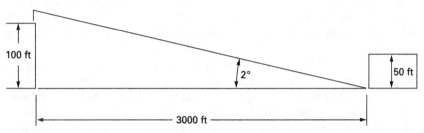

Figure 6.47 Downtilt.

mize the coverage of the cell site and achieve the necessary attenuation of the signal to facilitate frequency reuse.

However, downtilting an antenna to improve the performance of a network is not the only method that has been used. One example occurred when a cell site was at the bottom of a major hill and the road pass was of course several hundred feet and about a ½ km from the site itself. The situation involved major system problems where mobiles coming over the pass would either originate on a distant cell site, the common problem, or experience severe interference due to poor C/I for the area.

The solution at hand was to have a positive antenna inclination, of a couple of degrees, to ensure that the main lobe was almost pointed right at the pass itself. The simple objective was to ensure that there was the correct dominant server for the pass and that this individual sector would lead the mobiles into the rest of the system as they came down the major mountain into the rest of the system.

Another example involved the engineering department recommending tilting an antenna for just one sector of a six-sector site by 10°. An adjacent system operator complained of interference and the design of the day was to employ downtilt. The problem later found was that the real nature of the problem was never indexed and the wrong cell was downtilted. However, the engineering department was adamant that the implemented fix was the correct method.

The end result was that the sector downtilted had its entire antenna inclination removed so no downtilt was employed and a simple retune of two cell sites was implemented to fix the problem. The retune resolved the problem because it was adjacent channel interference and not cochannel interference. The lesson here was that everyone assumed what the problem was and never bothered to first investigate its true nature or contemplate that the fix implemented might not have worked.

When planning on altering an antenna system inclination it is strongly advised that a test plan or MOP be used. The key element for the plan which must be done prior to altering the antenna pattern for a cell site is to define what your objective is before you begin. Defining the objective beforehand seems simple in nature but it is imperative.

The following is a procedure for altering the tilt angle of an antenna system, usually a sector, for a cell site. The procedure here, as with other procedures, can and should be modified to reflect the particulars of the situation. It is meant for a three-sector (trisector) cell site. If you have a six-sector cell, a modification to this procedure is needed and involves simply increasing the number of data points to test. Altering the tilt angle for a cell site can involve increasing and decreasing the angle of inclination currently used there. Many improvements to a network's performance have been achieved simply by reducing the tilt angles currently employed at a cell site.

Downtilt procedure

1. *Identify the problem.* There are several methods for helping identify the problem itself. Some of the methods that you should use are statistics on the primary and surrounding cell sites. The expected propagation of the site should be checked on the computer molding system you employ. A key concept here is that the solution at hand might be to do nothing, an often overlooked option. During this phase of the procedure it is important to identify any potential coverage problems you might have for the site and historic problems to prevent the proverbial reinventing of the wheel syndrome.

2. *Operations check.* Before any additional work is performed toward altering the site it is important to have operations validate that there are no major problems with the site's performance which might be the root cause of the problem itself. Keep in mind that every cell has problems but the exact scope and nature of the problem at hand must be weighed against what the reported problem is.

3. *Physical observations.* The next important aspect of this procedure involves a physical visit to the location by the design engineer involved. The objective here is to help formulate the testing level that is needed. The engineer with operations personnel inspect the antenna system, if possible, to observe or confirm any problems identified previously. Sometimes the records that are kept on the site do not reflect what is there physically. At this stage of the process the antenna system should be checked for orientation correctness. Also an S11 test should be performed on the antenna system to ensure there are no additional problems with the site.

4. *Drive testing.* The next step in the process is to perform a drive test of the area in question. The drive test that is done needs to either support or refute the basic premise of the problem you have or that someone else has identified. Drive testing of the area involves signal strength and call processing testing and physical observations made of the site and the surrounding terrain, including buildings. During this process the basic premise of what you intend on doing for this site is identified. If the test involves trying to reduce the signal level from the primary cell at a specified location, the testing that needs to take place should reflect this. The drive test needs to be conveyed on a map using the format and methodology mentioned in Sec. 6.6.

5. *Define cell coverage area.* Based on the data collected and the desired system configuration establish the desired coverage as compared to what it is. Establish the desired coverage area of the site, taking into account terrain, adjacent cell, future system expansion, and frequency planning concerns.

6. *Design review.* This part of the procedure involves putting together a plan of action based on the results achieved so far in the previous steps. The design review should be achieved prior to any further action on the site. It is important at this stage of the process to ensure operations is directly involved with the activities planned. The items to be reviewed at the design review meeting are (*a*) the objective for the antenna alteration, (*b*) desired coverage area for cell site and sectors, (*c*) preliminary tilt angles based on items *a* and *b*, (*d*) pre- and posttest plan, and (*e*) pass/fail criteria.

7. *Test plan.* For this portion of the procedure a multistep sequence is described that has been successful time and time again. The test plan involves collecting data at various predefined test points and then altering the tilt angles to achieve the desired results. The test procedure described below involves measuring the forward and reverse paths for a cell site in establishing the tilt angle using a spectrum analyzer or service monitor at a cell site for measuring the reverse path energy. A mobile, with RSSI capability and calibrated, is

used to measure the forward energy of the cell site and also to send energy back on the reverse path.

Data points are picked in advance and the mobile is then moved to each of the data points during the testing sequence. Usually six data points are used for a three-sector cell site. They are set up so they fall directly in the center of the sectors and at the sector boundaries for the cell. Figure 6.48 shows an example of the data point locations for a three-sector cell. Having a predefined signal level for the border of the cell site, defined by the design criteria, the antenna tilt angles are adjusted to meet the objective. Based on the tilt angles used it might be necessary to adjust the adjacent sector to prevent one sector from dominating another because of the tilt angle employed.

Forward and reverse energy readings are collected for each of the data points selected. The values collected are then compared with the desired results anticipated to ensure that they are achievable. The antenna system then has the tilt angles employed based on the preliminary design calculation set forth in the design review. The forward and reverse energy is then measured at the data points again. If required, additional antenna tilt angle alterations are made to achieve the desired objective for the site.

The next step is to conduct a call processing test to verify that the cell is performing as desired. The call processing test involves driving the same routes utilized previously as part of the benchmarking process. Statistics analysis also is done at this stage to verify that no adverse effects have occurred to the region as a result of the actions taken.

8. *Closure.* The last step in the process is to issue a closing report detailing the results of the activity. The report needs to include a before

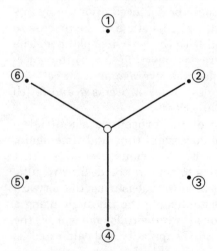

Figure 6.48 Downtilt test locations.

and after evaluation that involves performance metrics. The net result for the closure report is to determine if additional action is required regarding this site. A copy of the final report needs to be put into the site file records to ensure the documentation trail is completed.

6.12 EMF

One design criterion that needs to be factored into the design and cell site acquisition phase is electromagnetic radiation compliance (EMF). Whenever this topic is raised it invokes a wide range of responses from the sublime to outright panic from both the community and the wireless operators. Almost everyone who has been in the wireless telecommunication industry, primarily cellular, has been asked if electromagnetic radiation is safe. The answer to the EMF safety question is yes, the electromagnetic energy emitted from the site is safe, but there are a few conditions. The but, or any other caveat, to the electromagnetic radiation question is the part that tends to draw the focus of attention since a definitive yes or no is the real desired response.

I will not try to evaluate the multitude of technical studies already done or currently being conducted. However, I will discuss the various issues that pertain to all current and future operators for constructing and maintaining a communications site. Most current operators tend to take a reactive role in dealing with the EMF topic. Specifically the unwritten rule tends to be: Do not discuss it and it might go away. Unfortunately or fortunately, depending on your point of view, this issue is not going to disappear but will most likely intensify with the new PCS spectrum that has just been made available.

During the site acquisition phase many things can be accomplished in the beginning steps to minimize the controversy associated with electromagnetic fields and the installation. Currently towers or monopoles tend to draw the most responses, but more structures in the urban and suburban areas are being located on top of existing building structures. Placing sites away from schools and residential homes is one of the techniques that has been pursued to help build out networks and avoid the controversy. However, this leads to two primary problems, the first leading to many system design compromises, since there is no real grid and no perfect site. The other issue ties to the emotion controversy: If it is safe, why are you avoiding these areas for building sites? The answer which most operators and builders do not really mention is that locations are chosen that will avoid a controversy to help ensure that the site will be constructed within that year's goals and objectives.

Since speed is the primary driving force in building out a network, properly conducting the real estate acquisition phase can expedite the build program through selecting locations that do not require any variances prior to pulling the permit. However, no matter how well you perform your site acquisition process you will need to apply for variances at some of your locations.

Everyone who has had the pleasure of participating in the site acquisition process can tell you that several key items constantly arise, after the lease negotiation phase, are aesthetics, property devaluation, and electromagnetic radiation. The first two items are probably the easiest to address since they are the most tangible. However, the third item, electromagnetic radiation, will always invoke an emotional response from either the landlord, local residents, the board members, or all of them, since it is more times than not associated with cancer.

Prequalifying a potential site is one of the most important aspects for helping move the process along. Explaining to the landlord in advance what your intentions are prior to conducting a transmitter test is essential for determining if you will eventually receive flak from the landlord. Problems with the landlord have at times eliminated sites from ever being constructed after they passed all the design acceptance phases and a large amount of capital dollars have been expended. Some of the problems experienced are caused by the landlord and others are self-inflected during the initial site-selection phase because of miscommunication or deliberately leaving out key items.

In one example either the potential landlord did not understand what was being proposed or the communication system employee did not say that there would be antennas at the location. Another common example is that the residents at the location raised concerns regarding the EMF issue and the landlord decided not to pursue having the wireless operator as a tenant. In a third example the landlord understands exactly what is going on but when residents raise concerns it places all the burden on the communication company.

One of the key items for prequalifying a site that is on top of a building involves walking the rooftop, where applicable, and prior to any transmitter test determining where the potential antennas would be placed. By conducting a pretransmitter survey of the location you can avoid having the antennas placed right above someone's balcony or roof patio or next to or right at someone's window. If the design calls for a monopole or a tower to be constructed, having some flexibility in placing the structure on the property can also assist in the effort. More times than not a properly placed antenna or tower for a location can not only meet the design condition for the location but also minimize or eliminate any negative responses you can get for doing so. The simple rule is that with a little up-front planning and effort, less time, effort, and money will be spent later trying to convince everyone that it is all right to have built this installation.

One of the more expedient methods for site acquisition involves colocating with another service provider on a tower, monopole, or building. Primarily if a tower already exists it is a target of opportunity, provided the location meets the engineering design criteria established for that area. However, whether you are a sole occupant or one of many at a communication site a few rules must be followed that pertain to electromagnetic fields. The intermodulation issue is usually the primary focus of the RF engineer in a colocation situation. However, there is another critical issue that tends to play a background role in the site selection process. Currently cellular system operators utilize part 22 of the CFR 47 rules for determining the allowable electromagnetic emission levels from their facilities. PCS providers will be utilizing part 24 of the CFR 47 rules for determining the allowable electromagnetic emission levels. One key common point with both of these is the fact that they both reference part 1 of the CFR 47 with specific reference to 1.1307. The importance of the above items is that they both utilize C95.1 as the basic reference for determining the allowable electromagnetic field strength permitted, and this is required to be followed. Cellular systems currently utilize the C95.1-1982 specification while PCS utilizes C95.1-1992.

The C95.1-1992 specification has two basic sets of criteria that must be followed in determining if a communication facility meets the regulations. The two basic criteria pertain to a controlled and uncontrolled environment. The specification has a definition of what controlled and uncontrolled means, but the difference is that controlled pertains to the workers for the communication company and usually applies to its RF technicians. The uncontrolled specification pertains to everyone else, i.e., the general public. The primary difference between the two is that there is a difference in power levels a person is allowed to be exposed to. For example, as a general rule the uncontrolled environment is about one-fifth of the controlled environment. The C95.1-1982, however, does not differentiate between a controlled and an uncontrolled environment. Many cellular operators are already following C95.1-1992, if for no other reason than for public relations.

Table 6.1 demonstrates the difference between controlled and uncontrolled levels for the 800-MHz to 2.0-GHz band. At first glance you will notice that the specification not only differs for controlled and uncontrolled but also varies with frequency. Currently cellular base stations transmit between 869 and 894 MHz. The new PCS systems will now be operating between 1850 and 1990 MHz. There are several caveats with Table 6.1 when you reference specification C95.1,

TABLE 6.1   Power Density Chart, in mW/cm²

		Frequency, MHz				
	Equation	880	900	1800	1900	2000
C95.1-1982	f/300	2.933	3.0	6.0	6.333	6.666
C95.1-1992:						
Controlled	f/300	2.933	3.0	6.0	6.333	6.666
Uncontrolled	f/1500	0.586	0.6	1.2	1.266	1.333

in particular the time averaging that this measurement is to take place over.

Note:

1. f is the particular frequency of concern in MHz.

2. Power density levels here are the maximum permissible exposure levels.

3. Values taken from MPE tables for controlled and uncontrolled environments (C95.1).

One other item that is rarely covered or discussed is the composite power issue associated with the communication site. Generally in this situation the particular company that is petitioning to get on the facility will point out that its emission levels are well below the specification levels and that there is no real issue here. When you are colocated with another operator, as will be more and more the situation, the total power from all the transmitters must be factored into the equation at that facility.

This last point is often not chased down by the operators because most of the colocation situations are run by a site manager. The degree of documentation from one facility to another varies greatly, and calculating the composite power for a facility becomes very difficult. The alternative is to take a spectrum analyzer and/or a wideband power density probe to measure power density in the area around the facility. This might meet with some success on the superficial level, but it still makes the overall determination of composite power difficult since full loading of cellular, paging, and two-way systems is not normal. However, conducting a field measurement is a viable method for determining what the overall power density levels are at the location. The biggest disadvantage to using the physical testing methodology is the time and effort spent collecting data when an analytical method could suffice for over 90 percent of the locations.

During the site acceptance phase of the project it is strongly advisable to keep the antennas away from a common access area in any residential buildings. Although this should be common sense, most of us can recall situations where this simple first step did not take place, resulting in much after-the-fact work and damage control. When you are conducting the initial site survey a simple chart with distances and power levels comparing them to Fig. 6.49 should be used as an initial step in the process. Once the actual antenna locations are decided upon during a structural review, the distance calculations could be checked again to ensure that everything is all right at this stage. I would suggest as an up-front process utilizing the free space path loss equation as the first round since it is more stringent and can be trimmed down later with an actual field measurement. The equation utilizes a 20 log scale which seriously overemphasizes the propagation characteristics. The free space path loss equation does not take into account the air interface loss, antenna pattern correction, dynamic power control, modulation techniques, or the fact that the site is not fully loaded all the time. A quick glance at Fig. 6.49 tells you that for a tower or monopole situation where the height is at least 100 ft above the ground there is no issue with the emission levels. However, in a rooftop situation with apartments or offices across the street or right below the antenna structure the situation could be different, especially in a colocation environment.

You can do a variety of things to keep in compliance with the specification if the calculations show you could possibly exceed the levels. One obvious method is to reduce the power, but this has a negative impact on the coverage area you are trying to satisfy with this site. Another suggestion is to relocate the antennas so you get more vertical isolation; however, this might lead to aesthetic problems, installation complications, and cost increases for the facility. A third suggestion is to lower the maximum number of channels available at this facility, but this would eventually make it a limiting site in your system design. A fourth suggestion is to utilize a wideband probe and a spectrum analyzer, conduct a series of measurements to document the actual levels that would be there, and make adjustment from there.

When the time comes to meet with the local board for requesting a variance it is important to address the EMF issue head on. There are two primary schools of thought on this topic, proactive and reactive. Most cellular communication companies promote the reactive method for dealing with EMF. This method has its advantages in that the philosophy is not to educate or notify the groups that will oppose the construction of the facility. The other philosophy is to be proactive in the site-acquisition phase, especially setting up an abutters' meeting and sending out preliminary information about the intentions with a

Cell:

Date:

Sector 1

# Channels	19
Setup Channel	1
ERP/Channel	100W
Total Power	2000W

Distance, ft	Total Power, W	Power Density, mW/cm^2	Max for Band, mW/cm	% Budget
10	2000	1.713998474	0.586666667	292
25	2000	0.274239756	0.586666667	47
100	2000	0.017139985	0.586666667	3

Figure 6.49 EMF power budget.

brief drawing prior to the mandatory mailing. Both of these philoso-phies have met with success and failure on a site-by-site basis. A proactive campaign is the desired method, with the amount of build-out activity that will take place with the new PCS entrants.

At a planning board or zoning board meeting the EMF issue will be raised more times than not. One common step that is overlooked dur-ing the process is that the presenters tend to focus on the board and not on the public. Specifically if the public does not want to have the site the chance that the board will approve it is low, even in the face of a costly appeal that the town will pay for. It is suggested that with regard to the EMF issue a form of "leave behind" be presented that contains all the topics that are normally asked for. You can either craft your own or rely on the information that the Cellular Telephone Industry Association and the Electromagnetic Energy Association have available, which is very useful and easy to acquire. However, crafting a handout for your individual company may make sense, but the internal reality of its taking place, after all the reviews, is not bright. Either method will generate some immediate benefit in that you are addressing the EMF issue up-front and people can leave with something in their hand.

One technique that has found merit in presentations to a board is to reference the emission levels to the C95.1-1992 specification. This method avoids trying to equate the power level of one service to that of another, which is a common practice. It has been found that this reduces the number of questions that are not relevant to the topic at hand, getting the variance approved. Several points are chosen with the aid of the real estate representative that the local board or resi-dents might be interested in knowing.

It is also strongly suggested that the EMF discussions be rehearsed in front of the management of the company. The testimony or com-ments issued by the employees during this time are recorded, and it is best not to find a surprise in there when generating an appeal. Having a common set of answers for routinely asked questions will ensure a uniform response from the company as a whole. That way when a subscriber or resident near a cell site calls the company to ask a question the same response is given from customer service, sales, and the technical staff. The most direct method of establishing a uni-form response to EMF questions is to refer to an EMF policy. A few companies have established or begun to establish an electromagnetic or EMF policy to be followed by their employees. Some have gone to the point of employing personal dosimeters on their technicians in their guidelines.

Once the site is on the air all that you will know about it deals with its performance and centa call seconds (CCS) or erlangs. However, questions and concerns do arise from residents who are near the sites and want to know how it is affecting them. There should be a phased approach to handling people who call in to customer service or through corporate communications requesting or demanding information. Specifically a standard package of information should be ready to send out to them that contains the handouts that you utilize for the planning and zoning board meetings with a standard cover letter attached. The second level of response should involve a brief conversation on the phone with the resident asking specifically what their concern is. Utilizing the standard answers that you use to prepare for zoning and planning board meetings you try to handle the situation as well as possible, with the objective of resolving the situation then and there. If this does not work, the next phase is to conduct a site visit and have a one-on-one with the individual, utilizing someone from customer service or corporate communication with engineering.

Most times the standard package or a phone conversation can handle the situation. However, in a few incidents an actual field visit utilizing measurement equipment takes place, which tends to heighten the anxiety level of the person involved. In the last situation you have entered into a no win scenario since the person only wants to hear no levels are present and even 1 percent of the standard is still 1 percent more than they want. A variety of mitigation solutions are available, besides those mentioned previously, for handling location-specific problems which are cost-effective. The key to implementing any mitigation technique is ease of installation and minimization of complexity.

References

1. American Radio Relay League, *The ARRL 1986 Handbook,* 63d ed., The American Radio Relay League, Newington, Conn., 1986.
2. *The ARRL Antenna Book,* 14th ed., The American Radio Relay League, Newington, Conn., 1984.
3. AT&T, *Engineering and Operations in the Bell System,* 2d ed., AT&T Bell Laboratories, Murray Hill, N.J., 1983.
4. Carlson, A. B., *Communication Systems,* 2d ed., McGraw-Hill, New York, 1975.
5. Carr, J. J., *Practical Antenna Handbook,* TAB Books, Blue Ridge Summit, Pa., 1989.
6. Code of Federal Regulations, CFR 47 Parts 1, 22, and 24.
7. DeGarmo, Canada, Sullivan, *Engineering Economy,* 6th ed., Macmillan, New York, 1979.
8. Dixon, *Spread Spectrum Systems,* 2d ed., Wiley, New York, 1984.
9. Hess, *Land-Mobile Radio System Engineering,* Artech House, Norwood, Mass., 1993.
10. IEEE C95.1-1991, IEEE Standard for Safety Levels with Respect to Human Exposure to Radio Frequency Electromagnetic Fields, 3 kHz to 300 GHz.
11. ITT, *Reference Data for Radio Engineers,* 6th ed., Howard W. Sams & Co., Inc., New York, 1983.
12. Jakes, W. C., *Microwave Mobile Communications,* IEEE Press, New York, 1974.
13. Johnson, R. C., and H. Jasik, *Antenna Engineering Handbook,* 2d ed., McGraw-Hill, New York, 1984.

14. Kaufman, M., and A. H. Seidman, *Handbook of Electronics Calculations,* 2d ed., McGraw-Hill, New York, 1988.
15. Lathi, *Modern Digital and Analog Communication Systems,* CBS College Printing, New York, 1983.
16. Lee, W. C. Y., *Mobile Cellular Telecommunications Systems,* McGraw-Hill, New York, 1989.
17. Lee, W. C. Y., "CoChannel Interference Reduction by Using a Notch in Tilted Antenna Pattern," IEEE, 1985.
18. Lee, W. C. Y., "Effects on Correlation between Two Mobile Radio Base-Station Antennas," IEEE, 1972.
19. Yarbrough, *Electrical Engineering Reference Manual,* 5th ed., Professional Publications, Inc., Belmont, Calif., 1990.

Network Performance Measurement, Optimization, and Troubleshooting

Once a system is operating in a live environment its performance will need to be constantly measured and optimized. In addition, as service problems arise troubleshooting will have to be conducted using various techniques as they pertain to a wireless network. The topics discussed in this chapter provide a description of the major network performance metrics, ways to monitor and optimize these metrics, and procedures to troubleshoot call delivery problems.

7.1 Network Performance Measurement and Optimization

7.1.1 Network node performance measurement and optimization

Switch CPU loading. All switching equipment contains processors whether the node architecture is of a distributed or hierarchy design. These processors are usually divided between the main or central processor and the subordinate or regional processors it controls. The processor loads of a switch with a hierarchy-based design structure should take into account all levels of processors and their specific function in the delivery of a mobile telephone call. It should be determined which processors have the highest traffic levels and thus which are most susceptible to reaching an upper threshold and creating possible problems as the traffic in the system increases. A processor load study for a switch architecture based upon a central processor will be mainly concerned with measuring this processor's traffic load rather than any of the secondary processors it controls. However, the secondary processor loads should also be reviewed but not as frequently.

When the switch or node is in a stand-alone configuration, i.e., not in an application environment, a baseline (initial) load is present on each of its processors. This load consists of the basic administration and maintenance processes the CPU (central processor unit) and its subordinate processors perform. This value is a good indication of how well the switch vendor has designed the product. A typical baseline value may be in the 5 to 7 percent range. A 10 percent baseline load would be rather high. Obviously a higher baseline load means less processor capacity available to the application environment. Consult your vendor for this design information when performing your processor load analysis. Many other factors can add to the baseline load and overall load of the system switch processors, i.e., a new switch operating system, new system features or functions of the node, etc. The switch vendor should provide the percentage increase or decrease in the processor load for each of these network changes. Furthermore, the network engineering department working in conjunction with the vendor's support personnel should verify these changes after they have been implemented to assure that they meet the design specifications previously defined.

A switch vendor should also be able to supply you with the maximum operating limit of the processor with an acceptable variance and/or tolerance. When reviewing this value take into consideration that a processor cannot operate at 100 percent capacity for a number of reasons. First, as a processor reaches its maximum operating level it begins to shed tasks assigned to it for processing (see Table 7.1). These tasks are prioritized in levels of importance with regard to the operation of the switch. For instance, the administration and operational tasks of a switch may be assigned a greater weight and/or priority than the actual call processing tasks. This design is sometimes used by the manufacturer to assure the switch does not encounter a

TABLE 7.1   Example Switch Processing Task Prioritization

Switch function	Priority
I/O communication functions	Highest level
Call processing functions	Secondary level
System billing functions	Third level
System statistics collection functions	Fourth level

catastrophic failure whereby all communications and control of the node is lost. Thus the input and output (I/O) communication functions of the switch are assigned the highest priority over the other functions of the node. The operation of the I/O functions is necessary and vital for the switch operators to troubleshoot and correct any system error or reduce the processor load, whichever caused the disruption. This type of switch functionality is required to quickly restore it to its normal operating condition. Further priority assignments may place the call processing tasks at a higher priority than the tasks used in the collection of the system operating statistics. The reasoning here is that the processing of mobile calls is more important than the collection of operational data for a short period of time. Once the loss of operational data is noticed it is assumed that corrective action would take place immediately without the loss of system revenue. Here is an indication that a system could be reaching its upper capacity (the loss of statistical data from the switch, especially during the system busy hour).

Another reason that a processor cannot operate at 100 percent capacity is the fluctuations in the traffic load on the switch and thus the processor itself. Given a sampling rate and a time in which to sample the processor capacity, the final load value then is an average of all the sample measurements. Obviously the greater the sample rate the more accurate the average value of the processor load. (Take note, if a switch allows the operator to set the sample rate for the processor load measurement be careful not to set this value too high or it may affect the call processing functionality of the switch itself.) The average value of the processor load measurements cannot detect all the high-traffic instances ("spikes") that occur during the sampling period. Therefore, an average processor load value or measurement might read 80 percent but there may very well have been a peak, or spike, of 95 percent. It is with these concepts in mind that the operational loads of a processor are set to a value less than 100 percent of its full capacity. For some switches this value is set at 75 percent and for others it is set at 95 percent.

The actual assignment of priority levels to the various tasks or functions in the switch and the level in which the processor load shedding begins are dependent upon the switch vendor and the design they chose for their product. While these switch design concepts are critical to the operation of your system, the environment in which it operates is equally important to the performance of the processor and the time it will reach its maximum load.

The methods used to measure the processor load of a switch vary from one manufacturer to another. For some vendors this procedure involves simply issuing a command to the switch for a specified time and then collecting the printed data and/or data file from the operations department. Other switch vendors, however, may require a collection program to be set up specifying the sample rate and the starting and ending times of the sampling period. Once the method of measurement is established, the CPU load data of the switch can be collected, tracked, and analyzed. It is recommended to measure this load during the system busy hour (typically between 4 and 5 P.M., 1600 to 1700 hours) on a daily basis. These data can then be averaged for the 10 highest traffic days of the month for graphing and monitoring purposes. See Table 7.2 and Fig. 7.1 for an example of this process.

Upon reviewing the graph in Fig. 7.1 it is evident that the maximum processor load for switch A will be reached by the system in late January 1996. Before this limit is reached, plans for a new switch (switch B) should begin to relieve this loading in a 6- to 10-month time frame. Either a new switch must be added to the network or various system parameters will require adjustment. The addition of a new switch to the network should follow the guidelines discussed in Chap. 5.

Using the guidelines discussed above, if switch B is cut into service at its scheduled target date, 2 months before the projected maximum switch load is reached (November 1995), then an off-loading of switch A is experienced. This off-loading is a direct result of the transfer of system traffic to a new switch. The amount of traffic to off-load is estimated by the network and RF engineers prior to the actual cut date. Once switch B has been in service the actual off-loading value should be determined and the amount of variation in their design checked for accuracy against their initial projection. In Fig. 7.1 switch A is off-loaded 23 percent (69.8 to 54.0 percent CPU load decrease) by switch B. In addition to the off-loading aspect of this project the CPU loading trends of switches A and B need to be determined and the projection process continued for a multiswitch network. The projection of CPU loads for switches A and B is shown in Fig. 7.1.

As a temporary solution to the CPU loading problem a number of different system parameters can be adjusted in the switch to alleviate this loading problem. For instance, if a system is using the mobile registration feature whereby all system mobiles signal to the network their location in regularly scheduled periods based upon a specified algorithm (this process is further described later in this chapter) then a set of parameters exists for use as a temporary solution. By increasing the registration interval the total number of mobile registrations (location signals) being sent to the switch will decrease over a given

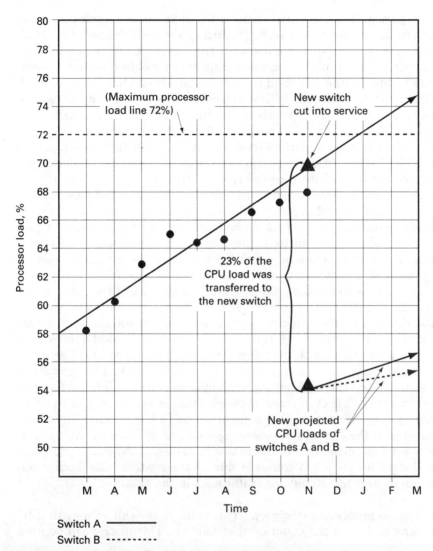

Figure 7.1 Switch CPU loading plot for market X (1995 to 1996).

time period. This, in turn, reduces the number of registrations the switch has to process, further reducing the switch processor load. However, this change comes with the network effect of reducing (slightly) the number of incoming calls to the mobiles, since the accuracy of the location function (mobile registrations) has been reduced. This is not as critical as it first appears, since the majority of calls in a cellular system are typically the mobile-to-land type (approximately 80 percent), but it is still a negative effect upon the network.

Again, this type of parameter adjustment is meant to be temporary until a final solution is prepared. Other parameter changes may include turning off the directed retry feature during the system busy hour in an attempt to reduce the number of redirected calls at this critical load period. Some vendors' switches perform maintenance functions continually over a 24-hour period while the switch is processing calls. These functions may be halted during the system busy hour, again as a temporary solution to relieve the processor load on the particular switch in trouble.

The purpose behind monitoring the performance of the processor is to plan and prevent ever having to get into a critical situation and utilizing one of the temporary solutions mentioned above. For this reason the CPU load of all the network switches should be monitored on a regular basis and after any new switch software's load or hardware upgrades have been completed in the network. This process is meant to keep the engineering staff informed of the CPU load and to determine the effect of these network changes on the switch's performance. It is also a good idea to monitor the CPU load after any major changes in the switch's database translations or if a new system feature is being introduced in the network. Table 7.2 shows the processor load data for switch A collected during the system busy hour, then averaged over the 10 highest traffic days in the month.

Switch call processing efficiency. This value is the call volume to CPU loading ratio (see the equation that follows). This metric can be used

TABLE 7.2   Switch Processor Load Data for Market x (1995)

Month	Processor load, %	Month	Processor load, %
Mar.	58.2	Aug.	64.8
Apr.	60.1	Sep.	66.2
May	63.0	Oct.	67.1
Jun.	65.3	Nov.	68.0
Jul.	64.2		
Vendor specified switch baseline processor load			10.0%
Vendor specified maximum switch processor load			72.0%

to monitor the call processing efficiency of the switch as well as indicate if a possible problem exists in the system. For example, a handoff border may need optimizing or a registration interval may need adjusting, etc.

$$\frac{\text{Switch call volume}}{\text{Switch CPU load}} = \text{switch processing efficiency}$$

The load of a processor is typically based upon the number of calls processed per second by the switch (the processor load traffic mentioned in the above discussions). However, a mobile call contains many components that affect the processor load on a per call basis. For instance, is the call from a roaming mobile or a home subscriber? Did the subscriber use a special feature? Was a handoff involved at any time during the call? All these functions contribute to the processor load for any given call. Therefore, a new feature introduced by your marketing department and utilized by your subscriber base will have an effect on the system. This addition to the existing cellular service may increase the processor load of the switch while the average number of mobile calls processed during the system busy hour would remain about the same. For this reason the load of the processors should be measured and plotted along with the number of calls processed per second by the switch during this same time period. These two metrics graphed together and compared are an excellent way to observe the call processing efficiency of a network switch.

Switch Processing Efficiency Example. Given that a system has two network switches, switches A and B, and switch A is processing 15 calls per second at a 50 percent processor load during the system busy hour while switch B is processing 15 calls per second at 62 percent processor load. The following observations can thus be made. Obviously switch B is the less efficient switch of the two. If the same system features are present on both switches and the traffic patterns, number of cells, number of trunks to PSTN, etc., are about the same, this difference may indicate a problem in that portion of the network. The performance of switch B should be analyzed in more depth, with possible optimization of its cells and global system parameters being conducted. For instance, if a high traffic area existed on switch B where the RF parameter settings of the cell sites were in error, this could affect the switch's efficiency along with other system variables such as voice channel blocking on the network facilities.

Note that the data for this plot should be taken from the daily system busy hour for both the processor load measurement and the number of calls processed per second measurement and not the monthly average data. Figure 7.2 is an example plot of these data for reference.

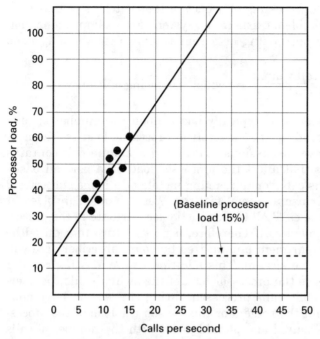

Figure 7.2 Switch processor load versus calls per second plot for switch X.

Switch and node downtime (service outage). Obviously, this is an important metric that needs to be monitored and tracked on a daily, monthly, and yearly basis to determine the rate of the occurrence, the duration of the occurrence, the cause of the occurrence, and what procedure and personnel were in place to correct the problem. If a system has multiple switching offices then compare the switch outage reports against the various mobile switching centers to determine where improvements can be made in personnel and operating procedures. Also, determine if the outage is due to errors in the operating procedures of the switch or if it is a quality issue with the switch vendor's software and/or hardware. This metric can also be used to track trends of declining or improving vendor support. Perhaps there is a scheduling problem with the number of network activities taking place in the system at a given time. Trying to perform too many complex and original projects in the network simultaneously or in a short time frame could cause many system service outages. Consult your operational or network management groups to address this performance issue in more detail. Figure 7.3 may assist in this effort.

System outage report for market "x"

Report date _____

Report number _____

Switch operator _____

Outage date _____

Outage start time _____

Outage end time _____

Description of outage _____

Switch data available _____

Personnel notification _____

Figure 7.3   Service outage example report.

Switch service circuit loading. The number of service circuits ordered and installed in your switch at the time it was cut into service was determined by the vendor's network engineers. The quantities of senders, receivers, conference cards, etc., is dependent upon the estimated traffic load expected to be processed by the switch. Most switches have a set number of these circuits installed as part of an initial switch configuration. However, as the system and the switch begin to carry more traffic the load on the processors as well as the service circuits increases. I can recall a case where a "live" switch stopped processing calls and was off the air because some of the service circuits were overloaded! There is no need for such a situation if these circuits are monitored on a regular basis and proper planning and ordering of new circuits is completed well in advance of an overload condition. A typical switch may include 200 multifrequency (MF) sender channels, 80 MF receiver channels, 80 dual tone MF receiver channels, and approximately 100 conference circuits. Take note that the use of these circuits is related to the use of system features and other services offered by your company. Any changes to the system service may affect the loading on the service circuits. Again, after any major system change or new service offering quickly monitor your system performance statistics for changes in performance and traffic load increases.

Switch and node total erlangs and calls volume. The total erlangs and calls carried by the switch during the system busy hour should be monitored to determine its percent traffic distribution in the system and to determine each node's call processing efficiency. This metric is also used to trend the loading of the individual nodes in the system. This is important to determine where the next critical loading limit will be encountered and when to begin plans to relieve this load. For example, if a network of four switches all carry the same percentage of the system traffic then as the call volume increases eventually all the switches will have to be off-loaded at the same point in time. This would be too large a task. Better to stagger the system load on the switches to in turn stagger the times each will need to be off-loaded. This spreads the planning and work needed for increasing system capacity over a larger time frame and thus makes it easier to manage.

Switch and node alarms. The list of current switch alarms is a good metric to track for monitoring the performance of an individual switch or an entire network. It also serves as a good indicator to determine if the node and corresponding interconnected network are operating in a clean and efficient manner. This list will contain the alarms that (although not necessarily critical) indicate minor problems with the switch database, cell sites, trunks, etc. These alarms might indicate data loaded in the switch that is being scheduled for use. For example, if a cell site is scheduled to be brought into service in a few days and the data have been loaded early in preparation for this cut then the switch may give some minor alarms of these devices being out of service. This is not a problem as long as the site *does indeed* get cut into service as scheduled! Otherwise these data could be left "lying around the switch" and cause unnecessary alarms *and CPU* loading. The same holds true for new data links and other network facilities that usually have their data loaded prior to being brought into service. Also, take note of any invalid or old mobile code ranges, devices out of service, problems with trunk groups, and other possible causes of network alarms.

Switch memory settings and utilization. Some switch manufacturers have the ability to dynamically set the amount of memory available for each function in the switch. While this is a great method to utilize the node's memory to the best extent possible, it also requires monitoring these memory assignments quite closely. Not allocating enough memory will reduce or even halt the particular function altogether. Such memory limitations may even cause a switch outage! To prevent this from occurring, monitor the memory utilization in your system switches and audit them on a regular basis to assure they have consistent setting in each node.

Switch timing source accuracy. Timing of the network nodes is critical for proper operation of the system. It is recommended to perform audits of the network timing sources on a regular basis (at least once a month) to assure they adhere to a stratum-2 level accuracy. Table 7.3 provides a listing of the specifications for the various stratum levels.

Auxiliary node performances. Monitor the performances of the network auxiliary nodes with the same level of detail at which the major switch nodes are monitored. They are an important part of your network and their performance should be reviewed on a regular but less frequent basis. Some auxiliary node examples are given below.

1. System DACs, multiplexers, and other dynamic transmission equipment used in the voice network.

2. STPs and other SS7 data network equipment such as monitors and DSUs.

3. Voice mail systems.

4. Network monitoring systems used to access and query the network switches and nodes. This includes network data routers, multiplexers, etc.

Node performance summary. The following is a summary of the major metrics used to measure the performance of a network node:

1. Measure the load on critical node processors.

2. Measure the processing efficiency of the node.

3. Track and monitor the outages of the node.

4. Measure the load on the switch service circuits.

5. Measure and report on the total erlangs and calls processed by the node during the system busy hour.

TABLE 7.3   Stratum Levels and Associated Timing Accuracy

Stratum level	Timing accuracy
Stratum-1	$\pm\ 1.0 \times 10^{-11}$
Stratum-2	$\pm\ 1.6 \times 10^{-8}$
Stratum-3	$\pm\ 4.6 \times 10^{-6}$
Stratum-4	$\pm\ 32.0 \times 10^{-6}$

6. Monitor the node alarms.

7. Track and monitor the node's memory utilization and capacity.

8. Check the node timing sources for a stratum-2 level accuracy.

7.1.2   Network link performance measurement and optimization

Network link performance.   When monitoring a network for data link performance the key facilities of interest are the links that interconnect the system switches and other major nodes. Cell site data links that do not utilize SS7-type data links, although important, will not be considered at this time. Critical metrics used in the measurement of a link or link set operating performance will be applied to the SS7-type signaling protocol. However, the concepts of these basic metrics can be applied to other transport-type protocols such as X.25.

Link traffic loading.   As discussed in Chap. 5, SS7 data links are typically designed to operate at a maximum 40 percent (0.4 erlang) utilization. Though the link can operate at a much higher level should another link in the link set experience trouble, the possibility of congestion exists. The traffic levels on an SS7 data link can be measured by using either the reporting system of the switch or separate monitoring equipment such as a protocol analyzer patched into the network in a nonintrusive (passive) manner. Some switches require (as in the case of the processor load data discussed above) a collection program to be written to collect the data necessary to calculate the traffic on a given data link in erlangs. The basic concept using this method is to determine the number of messages transmitted and received on a per second time interval. Next determine the number of bits per message and divide by the data rate of the link (56 kbits/s) to come up with the number of erlangs of traffic the link is carrying. Equations for measuring link performance:

(Number of MSUs transmitted) + (number of MSUs retransmitted) + (number of MSUs received) = number of messages carried per second

$$\frac{(\text{Messages per second}) \times (\text{bytes per message}) \times (8 \text{ bits per byte})}{56{,}000}$$

$$= \text{link traffic carried in erlangs}$$

These same data can be acquired in a much quicker and easier fashion from a network analyzer if set up properly. However, not all systems have the ability to patch into each network link and collect these data. Under these circumstances it is better to have the switch provide the data and have the network engineers process it along with the other remaining system performance statistics. If a data link is approaching a 40 percent utilization load on a regular basis, plan to add another link to the link set.

If a data link is showing a utilization rate of 90 percent or above and is causing service problems, first collect as much system data as possible and then reset the link and monitor the traffic levels once more. The reset may clear the problem. If it doesn't, a possible rerouting of the link may be necessary. Consult your network manager before performing any such actions and to receive the proper authority and consultation for conducting this work.

Similar to a node processor, an SS7 data link will shed (not transmit) lower-priority messages in the event of a link problem for the sake of maintaining the transmission and throughput of the higher-order maintenance messages. Four levels of priority are assigned to the SS7 messages for maintaining a data link performance. The lowest level (level 0) is assigned to normal SS7 messages and the highest level (level 4) is assigned to link messages (LSSUs, link status signaling units) used for maintaining and administering the link during conditions of congestion and errors.

Link retransmissions. Another metric used to measure the performance of a data link is its percentage of message retransmissions. A high number of retransmissions indicates a problem in the data network that requires further investigation. A typical retransmission level for a normal data link might be in the range of 0.1 to 0.4 percent. To calculate the percentage of retransmissions, use the following equation:

$$\frac{\text{Number of MSU octets retransmitted}}{\text{Total number of MSU octets transmitted}} \times 100$$

$$= \text{retransmission percent}$$

Link errors. The maintenance and administration of an SS7 data link will include the monitoring of the number of message signaling units received in error. This is a function of the SS7 protocol and should be monitored for evaluation of the link performance. Again, these data can be collected by one of two methods, either by a protocol analyzer or by the statistical collection functions of the switch itself. This metric should be recorded over 1-hour intervals for a typical operational day (approximately 7 A.M. to 8 P.M., 0700 to 2000 hours). As a guide,

400 signaling units in error over a 1-hour period would represent a nominal link performance.

The actual physical data link itself should be tested for an acceptable bit error rate prior to placing any data link into service. The ANSI specification is stated as follows:

>99.5 percent error-free seconds
[<432 ES/day at a DSO rate (64 kbits/s)]

This basically means that when a data link facility is ordered from the LEC or IXC it must be of a 64 kbits/s rate and operate 99.5 percent of the time without errors for a trial period of at least 24 hours before it is acceptable to be placed in service to carry live SS7 message traffic. Consult the ANSI Signaling System 7, MTP T1.111.2 (section 3) specification for more information about this test requirement.

Link changeovers. If the number of errors on a link exceeds the specified maximum limit, the link may be taken out of service and the traffic destined for that link will be rerouted to another link in the corresponding link set. Once the problem has been resolved the traffic will be rerouted back to the corrected data link and the routing of data messages will resume its original configuration. This change in the routing of the message traffic and status of the links in the link set is termed a "changeover." Obviously a large number of such events is a problem in the network and requires further investigation. As a guide, four changeovers in a 1-hour period would be an acceptable level of performance.

Link active time. A data link may experience brief outages during its normal operation in the network. However, excessive downtimes for a data link mean that the message traffic assigned for routing over this facility has been reassigned to another link while it attempts to resolve the problem. Initially this does not appear to be a problem since there is sufficient capacity on the other links in the link set. However, this type of network management will lead to further troubles if left unattended. In this particular situation the other links must carry more traffic than their normal intended design levels. Under these circumstances the capacity of the entire link set has been reduced. Should another link experience a problem, the entire set (containing two links) may need to be taken out of service. Even on a temporary basis this type of message rerouting will begin to cascade and get out of control if not resolved in a timely manner. As a basic guideline, link performance problems should be resolved within a regular operational week. This metric should also determine if a link outage performance is steadily increasing, decreasing, or remaining constant.

Link performance summary. The following is a summary of some basic metrics and guideline values for measuring the performance of an SS7 data link, link set, and corresponding network facilities. Many other parameters and methods are used to conduct monitoring and performance evaluations of this type. The reader is advised to consult the references listed for this chapter to obtain additional background regarding these topics.

1. Provide the basic link and link set definition for identification purposes.

2. Measure the actual traffic on the individual links for a specified time period (typically the system busy hour) by either collecting the number of MSUs (message signaling units) received and transmitted and using the equations provided or using a protocol analyzer for a direct reading of these data. The traffic levels for the link set need to be determined as well. The link set traffic load is simply the culmination of the traffic levels of the individual links that make up the set. Thus, if a link set has two links and each link carries 0.02 erlang of traffic, the entire link set carries 0.04 erlang.

3. Measure the number of retransmissions on a data link. As a guideline, an SS7 data link with a retransmission rate between 0.1 and 0.4 percent would represent a nominal performance.

4. Measure the number of message signal units received in error. If this value exceeds a given threshold (400) for a specified time interval (1 hour) then more in-depth monitoring and analysis of the link needs to be conducted. When commissioning a new data link or testing an existing facility check the bit error rate as specified by ANSI for a 24-hour given time period.

5. Measure the number of link changeovers in 1-hour intervals. As a guideline, four changeovers in a 1-hour period would represent a nominal performance.

6. Determine the overall service outage times of the link in terms of hour intervals. As a guideline, any link service outages should be resolved in one operational week.

7.1.3 Network routing performance monitoring and management

These metrics will apply to both the voice and data networks of a wireless system for measuring the routing efficiency and/or accuracy and for detecting possible problems in the network.

Routing efficiency (voice and data). Located in the switch is a subset of the database whose specific function is to analyze the dialed digits of the mobile subscribers when they place a call and to route these calls based upon the results of the analysis. This database subset is typically called the number translations of the system. Other names exist such as translations, B# analysis, etc. In spite of their many names these databases are basically designed to perform the same function, analyze the dialed digits, and route the call based upon the results of the analysis. This process applies to the voice network; however, a similar database is developed and operated to perform the routing of SS7 data packets within the network between nodes in much the same way that mobile calls get routed. The routing efficiency is therefore a measure of how accurately the call and data packet routing tables operate. First, let's discuss the routing efficiency of the mobile calls in a system.

The translation tables that are responsible for the analysis of the mobile dialed digits and the routing of the calls to their proper destination (the PSTN or another cell site for a mobile call termination) are developed by the switch vendor, the software engineers within the network engineering department, the marketing department, and the legal department. The last group is included since it is responsible for the company's tariff compliance which determines where and how calls can be routed outside the network to other telecommunication-type companies. The definition of the initial dialing plan (mobile dialing patterns) and any subsequent changes must be approved by the directors of both the marketing and network engineering departments. The actual implementation of the design, future alterations, and the maintenance of these tables is the sole responsibility of the software group. Any changes to these tables will immediately affect the call delivery service of the system mobile subscribers. So this is obviously an important design and operational aspect of the system.

In order to maintain these important tables, procedures should be put in place by the management to assure the accuracy of this database, to prevent unauthorized personnel from gaining access to these tables, and to provide a recent backup of the data in case a switch disaster occurs. With all this said, the purpose of this section is to provide a few basic methods to use in the measurement of the system routing efficiency and management of the switch translation tables.

Every switch has the ability to provide call delivery statistics to some degree. Although the forms of the data vary from switch to switch, a certain amount of data is basic to each manufacturer. For instance, the utilization of recorded announcements is a common statistic that should be readily obtained for review and analysis. The usage of such devices (actual magazines in the switch) will be of great benefit in measuring the routing accuracy of the translations and detecting changes in the calling patterns of your mobile subscribers. Every recorded announcement in the switch has a number associated with its message. This number should be included at the end of the recorded message along with some type of node identification. This will help immensely in the troubleshooting process described in the next section. Some typical recorded announcements are given in Table 7.4. With the utilization statistics for these recordings you can determine trends in the routing system. For example, if an abnormally high number of recording 02's are taking place then either there is a problem with one of the system trunk groups (voice facilities) or blocking is occurring on some of these facilities. It would be a good idea to have the software and traffic engineers review this problem and determine if it is a capacity issue with the network facilities or a routing problem that has entered the translations.

A large number of recording 03's may indicate a problem with the translations in regard to the node–mobile number range assignment and routing. If a new mobile code is loaded into a multiple-switch system and the routing is not set up properly, any subscribers assigned to one of these numbers will not receive service. Their mobile will not be validated and thus they will receive recording 03. By checking the routing for this mobile range the problem could be resolved completely or a step in the troubleshooting process reduced.

An example call delivery statistics report is shown in Fig. 7.4. By reviewing the report an engineer could look for trends and any large changes in the percentages of the system call mix and other call categories. Some useful categories for monitoring and improving the routing tables are the total calls completed percentage and the IMT (intermachine trunk) unavailable percentage. The total calls completed in a system will vary from month to month, but large percentage changes should not occur without some logical explanation. If no obvious net-

TABLE 7.4 Recorded Announcements

Message number	Call category	Recording
01	No page ack	"The mobile you have called does not answer. Please try your call later—message number A01."
02	Facility problem	"We're sorry your call cannot be completed at this time. Please try your call again—message number A02."
03	Unauthorized for use	"Your phone is not authorized for use at this time. For further assistance please call 611 from your phone and reference message number B03."

System call mix:	
Percent of total system calls (M-L):	80.0%
Percent of total system calls (L-M):	15.0%
Percent of total system calls (M-M):	5.0%
System calls by category:	

Category	Percent of total system calls
Completed calls	78.00
Unacknowledged mobiles	09.45
Invalid mobiles	00.05
Invalid ESN	00.10
RF channel unavailable	00.23
Land trunk unavailable	00.06
IMT unavailable	00.05

Figure 7.4 System call delivery statistics example report.

work changes have taken place recently, review the past month's translational changes for possible errors. It is a good idea to have a process established to archive recent translation and switch software load changes as a means to accomplish this review. Many software engineers store past changes on their computers at work. However, a more formal and defined method should be used that makes these data accessible to other members in the department as a backup in the event the originator is unable to respond to a problem.

Returning to our discussion, as mentioned above, the total calls completed is useful, as is the IMT unavailable category. Typically these facilities are designed with a 0.001 percent GOS (as discussed in Chap. 5). A value higher than this would indicate a blocking level on the system facilities higher than the design specified. Should this occur, have the software engineers review the network routing tables for errors. If no significant errors are found, consult with the network engineers to determine if the system IMTs need expanding because of an unexpected growth in the system traffic.

In addition to the mobile call routing tables the SS7 network routing tables need to be reviewed as well. Misrouting of SS7 data messages can lead to link congestion, message looping, and possible system service outages. These tables should be maintained much like the routing tables used in mobile call delivery. An overall review of the SS7 routing tables is recommended once a quarter. This type of review is especially important in a network with a large amount of SS7 network growth. As new links are added or reconfigured the routing becomes more complex and chances for error increase.

One method of checking the routing accuracy of the SS7 data network is to monitor the SS7 data links for occurrences of the various destination point codes (DPCs) or just point codes (PCs) of the network. The SS7 data messages routed over each individual system data link are defined by the SS7 routing tables. By viewing the distribution of SS7 messages (sorted by the DPCs) transmitted over a particular data link, an engineer can determine if the link is carrying message traffic that another network link could route more efficiently. As a network grows, many times a new link and/or links are available that provide a more direct route for the transmission of SS7 messages. The above method is also useful in detecting network messages with unknown DPCs. Messages of this type should be tracked for their source of origin and, if invalid, blocked before they are routed in the network.

These are just a few examples of how to utilize the data available in a network to optimize the routing of mobile calls in the system and to check the routing of SS7 messages in the data network. Other data useful in monitoring system mobile call delivery are available from your information systems department. Since they are responsible for the processing of the system billing records and producing the customers' bills they have access to a large quantity of mobile call delivery data. Consult the manager of this department for a listing of reports and data available to the company.

It is recommended that a system software design review and audit take place on a regular scheduled basis. Once a month for the mobile call delivery routing tables and, as stated above, once a quarter for the SS7 routing tables is a reasonable time frame to assure accuracy for this aspect of the network design.

Network routing performance summary. The following is a summary of steps used in measuring and maintaining the routing performance of the system translation tables:

1. Obtain a copy of the translation tables for reference.

2. Obtain statistics on the utilization of the system recorded announcements.

3. Obtain statistics on the network call delivery.

4. Monitor the utilization of the recorded announcements and the call delivery category percentages and note any unusual changes in the values.

5. Correlate large changes in the data to possible errors in the routing tables or the current network design.

6. Develop formal procedures to track and correct any errors that may develop in the routing tables. Also, keep a current copy of the

tables in a secure place at all times and limit access to the tables to experienced authorized personnel.

7.1.4 Network software performance

Every node in the system has an operating system. This software provides the ability to operate the system nodes for call delivery, billing, maintenance, and administration of the equipment, etc. The actual performance of the operating system itself is often overlooked in a network performance evaluation. The following are some metrics to use to complete a basic evaluation.

The initial software load and subsequent loads of a switch typically require additional switch hardware to provide more memory capacity. The amount of memory needed for a particular software load is dependent upon the type of operating system design utilized by the vendor, the number of system features supported by the load and active in the system, etc. For some switch vendors this is not a critical issue; for others new software loads require a large increase in the amount of memory required. This can sometimes be expensive, so be aware of this issue when preparing the system budgetary input to the company.

Another critical issue to be aware of when reviewing a switch operating system is the amount of processor load increase that a new software release will introduce. Sometimes a new load will increase the CPU load by as much as 5 percent. This value depends upon which system features are used and the kind of system the switch or node is operating within. A new software release can also increase or decrease the call delivery completion rate and the call delivery times (actual time it takes for the call to complete). These metrics can be measured in a simple manner by performing call testing in the network using a standard call test plan and known drive test routes. The results should be compared to an already established baseline test (a call test conducted during normal or stable system times) for observing any abnormalities and determining if the new load caused any degradation in the service in the system.

7.1.5 Network performance (general data)

The following is a list of general system performance and/or design data that can be used as a measure of the system's operation performance and capacity. These data will also be helpful in conducting specific performance and growth studies of the network switches.

Cell site data

- Number of cell sites currently in service on a switch-by-switch basis
- Number of cell sites projected to be cut in service by the end of the year

Subscriber data

- Number of subscribers currently assigned in the switch
- Number of projected subscribers assigned by the end of the year
- Number of subscribers assigned with the voice mail feature
- Number of subscribers assigned with other network features (traffic information service, etc.)

System and switch traffic data

- Average system busy hour usage (erlangs) (10 high day average per month)
- Estimated network maximum traffic level
- Average switch busy hour usage on a switch-by-switch basis
- Estimated switch maximum traffic level
- The usage per subscriber (erlangs)
- Percentage of subscribers registered and active in the system during the busy hour
- Number of registration attempts on the system during the system busy hour
- The registration interval assignments on a switch-by-switch basis
- Number of call attempts in the system during the system busy hour
- Number of calls completed in the system during the system busy hour
- Number of blocked calls in the system during the system busy hour
- Number of dropped calls in the system during the system busy hour
- Number of intrasystem handoffs during the system busy hour
- Number of intersystem handoffs during the system busy hour

- Average call holding time (seconds)
- Average call setup time (seconds)
- Switch processor load on a switch-by-switch basis
- Switch efficiency percentages on a switch-by-switch basis
- Total number of voice channels in service on a switch-by-switch basis (both RF and land)

7.2   Network Call Delivery Troubleshooting

Resolving call delivery problems in a cellular environment can be a lengthy process. It takes time to review the customer's complaint, collect necessary data, begin working with the customer (if possible), and conduct call testing if necessary. In this section some basic troubleshooting procedures will be given as an aid to this task as well as some examples of actual call delivery problems resolved.

7.2.1   Network call delivery troubleshooting procedures

Troubleshooting procedures (initial steps). When presented with a network call delivery problem (and I have had many given to me in my years in the cellular industry) be aware. Take my word of caution. Do not immediately rush off and begin delving into a large and time-consuming call testing and analysis project. I suggest you take a few preliminary steps that may, in the end, save you a tremendous amount of time and aggravation. First, who is reporting the problem? Are they following the previously established trouble reporting and resolution process? It has always been challenging for me as a manager of a group of systems engineers to guide them through their normal work load while they are receiving calls from other departments in the company about mobile call delivery problems without following any set procedures. I've seen three or four engineers get separate calls from another department to resolve a call delivery problem for a single customer's complaint. All these calls were received at about the same time. No managers were notified and so each engineer began working on the problem separately. They stopped their current scheduled work to start troubleshooting the problem. After some time I got involved, only to find out that they were all working on the same problem but none of them knew what the other was doing. A few phone calls from another department stopped the work of an entire group in the engineering department! So I stress again, follow agreed-upon procedures to resolve customer call delivery complaints. Tell your engineers that if they receive a call to begin work or troubleshooting outside their normally scheduled projects they should redirect the call to their manager.

Another great assistance when troubleshooting a problem in the system is to have the person reporting the trouble provide a written description of the problem along with a date and the name of its originator. This will help in the troubleshooting process of the problem and to record the total hours of work utilized for reporting to the department director and to other required company personnel.

Next, is the call a legitimate call? In other words, does this call actually exist in the network? A number of times I have heard of a customer complaint about using a feature in the network only to find out that it was not available! Or if it was available its functionality was removed from the system for one reason or another. To stop this type of confusion from occurring it is recommended that a manual be developed and maintained of all the calls supported in the system for both the home subscribers and the various types of roaming subscribers. This listing of the calls and features supported for each customer type will be a great reference in performing call delivery troubleshooting and in conducting network hardware and software acceptance call testing.

Finally, find out what activity is currently taking place in the system for that day or week. Is a new software load being implemented? Have there been some recent changes in the translation tables? Was there a change in the network configuration? Was there a system outage? What activities have the operations department been conducting? Was there any recent changes to the validation systems in the network? Did a previous system that your company had a roaming agreement with cancel their contract? Consult your accounting and finance department on this last item and request that they notify the engineering department when any such changes do take place.

Troubleshooting procedures (first level). Note, before starting the troubleshooting process it is recommended to have the customer care department quantify the problem to determine the magnitude of the customer impact. This will assist in scheduling the troubleshooting work among the other scheduled engineering projects. Obviously if the problem is critical and there is a large service outage, this troubleshooting will take precedence over other scheduled engineering work.

There are a number of levels to troubleshooting call delivery problems in a cellular network. The first level begins with the customer service department. Here the customer trouble tickets are generated and basic problem resolution is conducted. The most common problems are an incorrect subscriber profile with an invalid mobile phone number to electronic serial number (ESN) assignment in the switch, a mobile code that may not have been properly activated (code management problem), a customer who may not have the features assigned to them that they assumed, a customer who does not know how to use the cellular service and is dialing improperly, etc.

Many of these problems can be resolved by noting what call treatment was received by the customer. A call treatment is the classification of the various types of call terminations, for instance, a call completed to the intended called party that would be classified as successful call termination. A call that is sent to a busy recording while the called party was on another phone call would be classified as a busy call treatment. The call treatments are a very useful tool in troubleshooting call delivery problems. Each call treatment should include a switch code at the end of the recording. This code designates which switch in the network provided the call treatment and thus where the call was terminated. This is very important data!

Table 7.5 gives some basic troubleshooting tips. The message number at the end of all recorded announcements is the switch description followed by the recorded announcement number; i.e., message number A01 would represent switch A in the network and 01 would represent message 01 within switch A.

Troubleshooting procedures (second level)

If the call delivery problem cannot be resolved at the first (customer service) level, the engineering department will have to get involved and begin the second level of troubleshooting. This will involve the initial steps I mentioned above as well as call testing mentioned below. The first task after performing these steps is to verify that the mobile is indeed registering in the system (provided that the system is using autonomous registration). Most systems in service today use this function to locate the mobile within the system. All mobiles must first signal the system to be validated for placing and receiving *both* outgoing calls and incoming calls. By successfully completing this step the mobile has been able to communicate with the system and is ready to provide call delivery. If a mobile is not registering, check the profile of the subscriber in the switch. If these data are correct, check the mobile and RF environment to assure these areas are set up properly. For more information on the mobile registration process refer to EIA/TIA specification 553.

Once the mobile is registering, begin call testing to try to uncover the call delivery problem. Again, most times it is a switch database error that needs correcting or a problem with the validation of the

TABLE 7.5   System X Call Treatments and Possible Causes

Call treatment	Causes

Recording 01: "The mobile customer you have called does not answer. Please try to call later—message number A01."	a. No page response
	    No mobile answer
	    Mobile is outside of the service area
	    Mobile is turned off and does not have the phone transferred
	    Mobile is nonregistering
	b. Mobile may have an incorrect station class mark
	c. Possible system problem

Recording 02: "The mobile you have called has been temporarily disconnected—message number A02."	a. The customer is not active in the system (voluntary cancellation) b. The customer is not active in the system (involuntary cancellation)
Recording 03: "We're sorry your call cannot be completed as dialed. Please check the number and dial again—message number A03."	a. The customer has dialed improperly
Recording 04: "You are not allowed calls to this number. For more assistance please call our customer service number at 611 and reference message number A04."	a. The customer has calling restrictions on their profile such as local dialing only or international calling blocked. Or the called number may be blocked by the switch (i.e., ·900 or 976 numbers, 0 + dialing, etc.)
Recording 05: "We're sorry your call cannot be completed at this time. Please try to call later—message number A05."	a. All circuits are busy b. System problem
Recording 06: "Your mobile phone is not authorized for service at this time. For additional assistance please call customer service at 611 and reference message number A06."	a. Customer has an ESN mismatch in their profile b. Customer's ESN is on the deny list c. Roamer's NPA/NXX (mobile code) is not active in the switch d. There is no roaming agreement with this cellular company

mobile. This last item (mobile validation) is becoming more predominant in call delivery problems owing to the increased use of more complex and sophisticated fraud-prevention systems. A listing and description of the systems in use in your network is a must when performing call delivery troubleshooting.

If the problem is further embedded in the system, more extensive call testing and tracing must be completed. A trace on the call will reveal a great deal of information such as the numbers dialed by the mobile unit, the results of the translations of the dialed digits by the switch, the route the call was sent to, and the final treatment of the call. You may want to have the network software engineers and your local vendor support team get involved at this point to try to resolve this problem quickly while all these resources are available.

After a problem has been solved, it is recommended that a log be kept of the problem and its correction. This will serve as a means to track the work completed and to use as reference material in future troubleshooting and teaching company personnel how to resolve problems on their own.

Troubleshooting call testing procedures.   To conduct test calls for troubleshooting the first step is to set up the test mobile. This task requires a mobile phone that is easily programmed, a well-charged battery and wall charger, a test phone number, and the ESN of the mobile. The next step is to build a subscriber profile for the test phone in the switch database. This will include activating the number in the database, assigning this number to the test mobile ESN, assigning system features to the test mobile, and assigning the category of service to the test mobile (examples are local service only, local and toll service, local plus toll plus international dialing service, etc.). Take note, it is a very good idea to obtain a copy of the switch subscriber commands along with an explanation of each field for use as a reference while conducting this call testing. Also helpful is an example of a previous test mobile profile that has already been set up in the switch to use as a template. The collection of useful switch commands and example subscriber profiles can (and should) be carried over to other types of applications. For instance, printouts collected from past call traces would serve as excellent reference material when performing this type of work at a later date.

Next, collect the customer trouble tickets from the customer service department. These reports will contain the data needed to test and possibly resolve the call delivery problem in question. A standard customer trouble ticket will have the customer's mobile phone number, their ESN, their category of service assignment, a listing of the features assigned to the customer, a description of the call delivery problem including the type of treatment received when attempting this call, and where this problem occurred. With these data you can assess the magnitude of the problem and begin the troubleshooting process.

Another task to perform before call testing can begin is to check with your accounting and finance department for any requirements they may have regarding this use of the system service. Actual call testing may take a few hours or it may take months. The billing for the air time of the test mobiles will still be produced and given to the accounting and finance department for processing. If they are not informed of the call testing taking place, they may consider these calls fraudulent and take the necessary action to block service to these mobiles. Therefore, I suggest that a formal procedure be set up

To: Accounting Finance Dept.

From: _____

Network Engineering Dept.

Date: _____

RE: Request for test mobile activation

Mobile test: _____

Mobile ESN/ESNs: _____

Purpose of call test: _____

Start date of test: _____

End date of test: _____

Additional comments:

Figure 7.5 Example accounting and finance form used in performing network call testing.

to specify the test numbers to be used in call testing, the duration of the call testing, the purpose of the call testing, and the name of the person in charge who is responsible for this testing. This information will assist the accounting department in tracking the system use for actual paying customers and the use generated by call testing from the engineering and operation departments (see Fig. 7.5).

Note, it is recommended to have terminal access to each network switch all in one location to perform call delivery troubleshooting. This access is necessary to change the subscriber profiles of the test mobiles, check on the status of the mobile activity in the switch, and perform call traces. The location of the call testing must also have good coverage by the system. In a system with multiple switches it is a good idea to build test cells programmed to different control channels and make them available for this type of work.

In this test configuration the test mobile can be programmed to scan for only one of the test cell control channels. This allows call testing to take place on all network switches at any given time at one switch location. Having these test cells and test mobiles available is

System switch: _____

Tester: _____

Date: _____

Test mobile: _____

Test mobile ESNs: _____

Type of test call: _____

Start time: _____

End time: _____

Duration of call: _____

Call treatment expected: _____

Call treatment received: _____

Additional comments:

Figure 7.6 Example test call log form.

important not only for conducting call delivery troubleshooting but also for performing call testing in regard to accepting new switch software loads and bringing new network switches into service.

Another suggested requirement for performing network call testing is to have test call log sheets produced beforehand to assist in recording the test calls completed and their results. These data will be valuable input in the overall analysis of the call delivery problem (see Fig. 7.6).

Once the profile of the test mobile is built in the switch database and the other departments within the company have been given proper notification, call testing can begin. The next step is to check the test phone to make sure it operates properly in the system. Check to see if the phone registers, can make outgoing and receive incoming calls, and can activate and deactivate the features it has been assigned. If there is a problem with performing these functions, either the phone is not set up properly or there is a problem with the system itself and further data must be collected.

If the test mobile works properly, call testing can begin. If the problem is a customer complaint, review the trouble ticket from this customer and try to mimic the exact call by having the test mobile register or obtain service from the same network switch with which the customer claims to have a problem. Then try to make the same type of call the customer is having problems with. Take note of the type of call treatment you received when making the call. Did the call get completed? Did you receive the same treatment that the customer described in the trouble report? Did you receive a switch recorded announcement? etc. All these data are valuable when troubleshooting a system call delivery problem.

Call testing summary steps

1. Develop a testing lab complete with test cells and terminals assigned to every switch in the network for use in troubleshooting and completing network hardware and software acceptance call testing.

2. Develop a test call log sheet.

3. Obtain the customer trouble tickets from the customer service department for determining the type of problem calls, the mobile number of the subscriber, the features assigned in the customer profile, and the quantity and magnitude of the customer complaints.

4. Obtain a mobile phone and an unassigned number in the switch for call testing. If possible choose a mobile phone that is easy to program and a test number that is in the same 100's block where the customer's number is located. [A 100's block is a group of phone numbers clustered by the last three digits (0–99) of a 10 K range; i.e., 201-555-1200 to 201-555-1299 would represent a 100's block of phone numbers.]

5. Set up the subscriber profile of the test mobile in the switch database. Specify the type of service the mobile is to have and the features needed for testing.

6. Notify the other departments in the company of this call testing by using standardized forms and formal agreed-upon procedures.

7. Program the mobile phone for operating on the designated network switch. Check the phone for proper operation.

8. Obtain, for reference, a listing of all the recordings on the network switches and a listing of the mobile codes loaded and/or stored in the system. The list of mobile codes should contain the corresponding node designations where these codes are actually located. Obtain a list of all the service categories and their definitions as they are used in your system. Finally, a list of relevant and frequently used switch commands along with a set of example switch printouts would serve well as quick reference material, further expediting the troubleshooting process.

9. Begin call testing by trying to mimic the type of mobile call in question. Record your findings on the test call log sheets.

Network call delivery troubleshooting summary. The following list is a summary of the troubleshooting steps discussed above.

First-level steps (customer service):

- Is the customer service problem resolution completed?
- Is the customer trouble ticket completed?
- Are problems quantified and totaled for identifying the magnitude of the problem?
- Is the customer's profile available for use in further troubleshooting?

Network engineering call delivery troubleshooting initial steps:

- Don't rush!
- Who is reporting the problem?
- Are other company personnel working on the problem?
- Are proper problem resolution procedures being followed?
- Is there a typewritten description of the problem?
- Is this a legitimate call?
- What other activities are occurring in the network?

Network engineering second-level steps:

- Complete first-level steps.
- Complete initial steps for network engineering.
- Conduct call testing and call tracing.
- Log and record the results of the problem resolution.

These are just some initial ideas to assist in correcting call delivery problems that may arise in your network. Obviously, these steps can be expanded and augmented with your own data and experiences to provide more accurate and effective troubleshooting procedures.

7.2.2 Network call delivery troubleshooting examples

Example 7.1: Network Timing Problem This problem actually occurred between two multiswitch cellular systems interconnected with direct voice and data links (see Fig. 7.7). The problem was given to me in the network engineering department after the customer service technicians could not get a final resolution to the complaint. The problem was stated:

There was an intermittent call delivery problem occurring for system A customers roaming in system B. During selected times calls to these customers would fail and receive an "out of car" recording as their call treatment. The problem was recorded as taking place during the system busy hour.

An intermittent trouble many times signifies a load-related problem in the

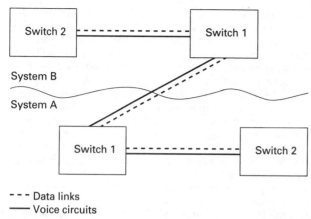

- - - Data links
——— Voice circuits

Figure 7.7   Example 7.1 system A and system B network configuration.

network. Somewhere a process was being overloaded or a timer was being expired. This is especially true since the customer complaints all indicated that the problem occurred during the system busy hour. I worked with system B to set up a test number and began call testing. After some time we were able to duplicate the problem. Upon review of the call trace in system B it was noted that the call timed out while waiting for the mobile acknowledgment to be returned.

Upon further discussion with system B it was brought to my attention that there had been a recent change in their network. System B had cut (brought into service) a new switch in their network (switch 2) and had changed the routing of system A's calls in the redesign. Instead of routing the calls directly to each of their switches as they were previously doing they were now using one of their switches as a tandem (switch 1). Thus a call to a system A mobile roaming in system B would first get routed to switch 1 for processing. The mobile would be paged, and if no acknowledgment was received the call was redirected to their switch 2. This additional processing caused the total duration of the call set up to exceed the maximum limit specified in system A's network, and so the call failed and was routed to an "out of car" recorded announcement.

Once we knew what the problem was, the solution was easy. Set up the original routing and conduct call testing. The routing was corrected and the problem ceased. So in summary, make sure the first-level troubleshooting is completed, acquire all necessary data and recent network changes, conduct call testing, perform a call trace to get more detailed information about the call problem, and finally test and record the solution.

Note, I have seen a similar problem where a number of mobiles were not being properly validated owing to an error in the network SS7 routing tables. The problem was occurring for subscribers stored in the system's HLR (home location register) and would occur only during the system peak busy hour. The subscribers experiencing the problem were receiving an invalid mobile recording from the switch until the SS7 routing error was fixed and the timing was not longer an issue.

Example 7.2: System Code Management Problem   Errors in mobile code management are one of the most common problems in operating a cellular system and a leading cause in call delivery problems. This is due to the high subscriber growth, subscriber churn (turnover rate), the demand for more mobile codes, and the constant shuffling of these numbers within the network to accommodate this growth. Under these conditions the chance for errors to occur is tremendous. Most cellular companies have a person or group dedicated to the task of managing the mobile codes in their network. Many names have been given to this team including the directory inventory group and the telephone number inventory group. The code management function and personnel are discussed further in Chap. 10. This group must monitor the usage of each mobile code (10,000's block of numbers provided by the LEC), order new codes based upon the remaining amount of unassigned phone numbers available and the expected subscriber growth provided by the marketing department, and keep track of where to store these codes in the cellular network. An important and time-consuming task, to say the least!

When customers first purchase a mobile phone, they are assigned a telephone number along with a category of cellular service they have chosen and any other system features that might prove useful. All this information is important, but perhaps the mobile number assignment is the most critical from a customer and system standpoint for the following reasons. When a mobile code (usually a 10 K block of numbers) is loaded in the system it is assigned a location in one of the network nodes as well as a serving central office from the LEC. If an error should occur in any part of this assignment these numbers cannot receive service. Now, it is highly unlikely that all 10,000 will be assigned to customers, but you can see the effect of an error in this part of the network. Therefore, management of these numbers is vital in preventing any call delivery problems from occurring to assure these resources are utilized in the most efficient manner possible.

Suppose a number of mobile codes (some partial and some whole) are moved from a switch to an HLR in a multinode system and an error occurs in the switch database. The error was made when the location of one of the partial codes was incorrect. The code is now rendered useless and all the mobile subscribers assigned to any numbers in that range will no longer be validated by the system and will be denied service. Typically these types of changes occur in the evening with only partial (random) call testing being completed to assure accuracy in the database. Then if the problem goes unnoticed the customer service department begins to receive calls from customers complaining about their service disruption. After a while a manager in this department will notice the trend and call the network engineering department to resolve the problem quickly. A quick review of the previous night's activity and database changes reveals the error, and the problem gets resolved. This type of problem is especially prevalent in call delivery scenarios involving mobiles roaming in other markets where the difficulty of managing codes is further increased between two companies.

Example 7.3: Network SS7 Problem   Since many of the cellular systems in North America use Signaling System 7 to communicate with each other for performing mobile call delivery, any errors that take place in this data network tend to have an immediate and large impact on the customer base. With the development and use of new network protocols, the need for trained and experienced network personnel, and the frequent network changes (both hardware and software) the chances of an error occurring in the SS7 network are great. Although this type of network has exceptional error detection and recovery, functionality errors still occur in the network and cause call delivery problems.

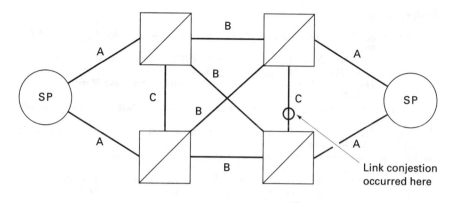

◩ = STP   A = access links
                 B = bridge links
◯ = SP   C = crossover links

Figure 7.8   Example 7.3 SS7 network.

Another actual call delivery problem took place when an SS7 routing error occurred in an SP (signaling point) and caused messages to begin looping around the network. The message looping caused the constant retransmission of the messages between two STPs. This, in turn, caused the higher-level link management functions to begin to intervene and start transmitting messages across the interconnecting C links in an effort to correct this problem. This caused a halt in the transmission of the lower-priority messages (MSUs) used for call delivery, which in turn stopped service to mobiles from other systems roaming in this network. (See Fig. 7.8 for more details.)

By placing a protocol analyzer on the link set and viewing its performance statistics it became clear what the problem was. The traffic levels on the C-type links within the link set were in the high 90's (almost at 100 percent congestion). This was causing the messages used for call delivery to get blocked and thus calls to mobiles utilizing these facilities were stopped. To correct this problem a sample data file was collected of the SS7 messaging on the links for later analysis, and the link set was then reset. The reset cleared the problem temporarily (1 to 2 days) and upon analysis of the SS7 message data captured the looping problem was discovered and fixed. This type of problem might have been avoided if the SS7 data network performance statistics were reviewed more closely and on a more frequent basis.

These are just a few examples of some call delivery problems that were resolved within a wireless network. It would be very productive to assign the engineering staff the task of documenting any and all network problems (not just call delivery problems) resolved for use in teaching new and current engineers how to troubleshoot and fix problems in wireless networks. This effort would assist the engineering staff in their understanding of the network's overall operation.

References

1. ANSI, 1988, *Signalling System Number 7 (SS7)—Message Transfer Part (MTP)*, ANSI T1.111-1988.
2. ANSI, 1988, *Signalling System Number 7 (SS7)—Operations, Maintenance and Administration Part (OMAP)*, ANSI T1.116-1990.
3. ANSI, 1990, *Signalling System Number 7 (SS7)—Monitoring and Measurements for Networks*, ANSI T1.115-1990.
4. AT&T Bell Laboratories, *Engineering and Operations in the Bell System*, Murray Hill, N.J., 1984.
5. Bellcore, *Bell Communications Research Specification of Signaling System*, vol. 1, no. 7, TR-NWT-000246, Issue 2, June 1991.
6. Bellcore, *Bell Communications Research Specification of Signaling System*, vol. 2, no. 7, TR-NWT-000246, Issue 2, June 1991.
7. Bellcore, *Digital Network Synchronization Plan*, GR-436-CORE, Piscataway, N.J.
8. Electronics Industries Association/Telecommunications Industry Association, Washington, D.C., April 1989, EIA/TIA-553.
9. Lee, William C. Y., *Mobile Cellular Telecommunications Systems*, McGraw-Hill, New York, 1989.
10. Motorola, Inc., *System Description Manual*, Motorola, Arlington Heights, Ill., 1987.

System Documentation
and Reports

This chapter discusses the various levels of documentation and system reports that need to be generated by both the network and RF engineering departments. The reports discussed in this chapter are recommended reports that should be used by every wireless operator. It is strongly suggested that you craft each of the recommended reports listed in this chapter to the particulars of your company.

A hierarchical approach to the system report generation is also presented in this chapter. The rationale behind a hierarchical report approach pertains to information requirements needed by different levels of management in an organization. Specifically a report that is used directly by a network engineer should be different from the report issued to a vice president for the same topic. While the hierarchical approach for report generation and dissemination seems straightforward, many reports go directly up the chain of command in a raw format.

The system reports presented in this chapter include network performance, RF performance, exception reports, customer care reports, construction status, operations, software configuration, and engineering project reports. It is suggested that each report have attached or immediately available a description of how it was produced. This should include the following as a minimum.

1. All the data contained in its report and the sources of the data

2. All the equations used in the calculations and the field in which they are applied

3. A listing of the software used in the processing of the data

It is also suggested that each report be checked for accuracy by running a set of simple values through the processing algorithm. The sample data run through the processing algorithm should also be checked against a manual run of the same data using a calculator or slide rule. I have caught some critical errors in the past by using this simple checking method.

There are some basic concepts to report generation and dissemination that you need to address for every report you are currently generating. When addressing any report or documentation process it is important to obtain the answers to the following seven issues:

1. What is its purpose?
2. Who will generate the report on an ongoing basis?
3. Who will act on the information that is in the report?
4. Who will receive the report?
5. Is the report needed?
6. What format should the report be in, i.e., electronic or paper?
7. How will the report be processed?

It is strongly recommended that you identify all the reports currently being generated in the various departments within engineering alone. A simple review of all the reports that are being generated will most likely result in several of the reports being identified as no longer needed. The review of the reports will also point to duplications in efforts within the organization itself. The duplication of various reports can be eliminated and resources can be better utilized to focus on other topics of direct interest to the company.

The reports themselves should utilize a document control number and be stored in a central location. The central location recommended for depositing all the engineering documents is the engineering library. It should contain all the meeting and/or project notes and all system reports. To facilitate access to the information the meeting and/or project notes along with all system reports should also be resident on the engineering LAN.

If an engineering library does not exist in your company it is strongly suggested that you develop one. The departmental administrator or a separate assistant should be placed in charge of the engineering library and should be responsible for overseeing the signing out and return of manuals, reports, etc. The administrator would also have the responsibility of updating the library and discarding outdated material. Examples of materials to have in the engineering library include industry specification, current vendor manuals, and reports.

It is as important to ensure that there is sufficient information in the engineering library as it is to prevent unwanted material from being stored there. The engineering library should not be a technical dumping ground for industry magazines, software manuals, and other items collected by the engineering staff. The success or failure of the engineering library is totally dependent upon continued vigilance for ensuring current and relevant information is stored in the facility.

8.1   Network Performance Reports

A multitude of network performance reports are required to be generated, disseminated, analyzed, and acted upon in a wireless system. The suggested reports listed in this section are generic in nature and need to be tailored to your individual requirements. The requirements need to be matched with your organization's goals and objectives plus any performance criteria that are deemed necessary. The proposed network performance reports are listed below.

8.1.1   Bouncing busy hour traffic report

The bouncing busy hour traffic report's intention is to identify the system and individual cell site busy hours. Through identifying system and individual cell site busy hours any network or local cell site blocking problems can be identified. The bouncing busy hour report should be generated on a biweekly and monthly level. The report needs to include weekend data as well and the weekday traffic information for a 24-hour period. This level of information should be disseminated only to the traffic and RF performance engineers for the network on a regular basis.

The system level report should be presented in both tabular and graphic method for ease of digestion. The information should be broken down to represent individual days of the month and compared to the previous year(s) data for trending information. The bouncing busy hour data for the system level should be the highest traffic volume period for any given day, over a 1-hour period, i.e., 17:00 to 18:00 hours. The recommended format is listed in Figs. 8.1, 8.2, and 8.3.

Figure 8.1 represents the comparison of daily usage versus the previous year's usage in addition to the system design level. The chart should be used to help map the general direction of the network in terms of usage. Figure 8.2 represents a daily system peak analysis of the network. The objective behind using this chart is to determine if there is a fundamental change in the usage patterns for the network. In the chart there should be some delineation between weekday and

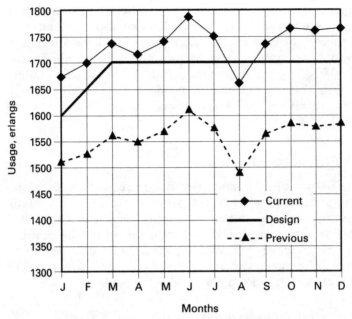

Figure 8.1 Monthly bouncing busy hour report.

weekend traffic usage. The data used for the daily system analysis need to be an average of the monthly data using the same data derived in Fig. 8.1. In both Figs. 8.1 and 8.2 the system traffic design level should represent what the system is designed to handle, on a macro scale.

Figure 8.3 is a table that represents the individual cell site and sector loading issues. The data need to be the bouncing busy hour of each cell site. The data in the table are straightforward and require little explanation. However, it is not recommended that any averaging method be utilized for this particular report. Instead the individual bouncing busy hour data for each cell and/or sector for the entire month need to be reported.

The report in Figs. 8.1 to 8.3 should be disseminated to the network and RF performance engineers only on a biweekly basis. The same report should be distributed to the manager for network and RF performance on a monthly basis. It should not be distributed to higher levels of management, however, since it is really meant for working level use. The bouncing busy hour report would be utilized, for example, in the redistribution of RF channels based on changing traffic patterns in the network. One might "pull" or remove channels from an otherwise underutilized sector during one season only to have to add channels back into the same sector during the next season.

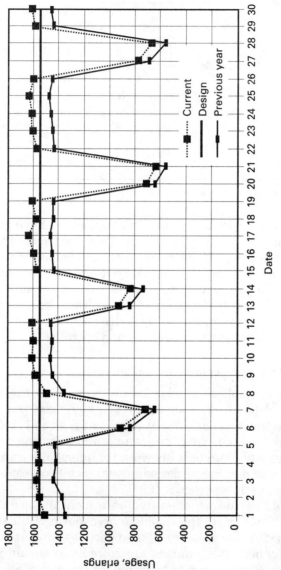

Figure 8.2 System bouncing busy hour usage.

Cell Bouncing Busy Hour Report

System Name: _____

Date: _____

Cell	Time	BBH Traffic, erlangs	Design usage	Radios assigned	Radios out of service	Blocking,%	Radios needed
1A	1600	5.3	5.84	11	0	1.2	11
1B	1700	6.1	5.08	10	0	4	12
1C	1700	1.2	1.66	5	1	3	5
2A	1700	1.5	1.66	5	0	1.5	5
2B	1800	3.1	3.66	8	0	1.0	8
2C	1700	2.3	2.28	6	0	2.5	7

Figure 8.3 Cell bouncing busy hour report.

If the traffic engineer in one situation would have reviewed the bouncing busy hour report for the past year, the trend in traffic would be noticed and the physical removal of RF channels would not take place. The correct alternative would be to leave the physical channels in their current place, remove them from service, and secure more radios for the vendor. The key concept to always keep in mind is that RF channel reallocation is a costly endeavor to the company in terms of dollars and labor from engineering and operations.

8.1.2 Telephone number inventory report

The use of the telephone number inventory report is vital for monitoring mobile telephone number usage, i.e., how many actual numbers are assigned to a system subscriber base at a given time. The report also is used for predicting, projecting, and trending the directory inventory growth. Through knowing the telephone number growth patterns an engineer can order additional mobile codes for future use from the local exchange carrier (LEC).

This report should be generated on a monthly basis and should indicate as a minimum the complete breakdown of all the directory numbers used in the network. It also should include the actual network nodes where the numbers are stored and should indicate the status of future codes soon to be introduced in the network and when they are expected to be released. Additionally the report needs to include the actual central offices (COs) out of which the codes are served to ensure proper planning and expediting troubleshooting.

The telephone number inventory report should be distributed to the network engineer responsible for the inventory tracking and forecasting, the network engineering manager, and the director of engineering. A summarized report should be issued to the external department responsible for allocating mobile numbers to wholesale and retail distribution points. An example of the telephone number inventory report is shown in Fig. 8.4.

8.1.3 Facility usage and traffic report

The facility usage report's purpose is to track the various interconnect facility usage levels in the network. It should be issued on a monthly basis and used for many issues associated with improving the network's facility performance. The facility usage report should be used to determine the best locations for the point of presence (POP) locations in the network. The selection of the POP locations, if done properly, will minimize the network infrastructure cost for delivering calls.

The report should also be used for verifying that the interconnect bills received for operating the network are valid. The interconnect bills should be reconciled against the facility usage report to ensure that nothing out of the ordinary is being reported or billed. The reconciliation of the facility usage bill is normally the responsibility of a revenue assurance department.

The facility usage report should be distributed to the network facilities engineer responsible for facility usage dimensioning, the network manager, and the director of engineering. The director of engineering should forward a sanitized version of the facility usage report to upper management with the interconnect bills. Figure 8.5 is an example of how a facility usage report should look.

8.1.4 Facilities interconnect report (data)

The facilities interconnect report (data) is intended to display the current configuration and performance of the network data links for detecting and resolving data problems and to plan for the dimensioning of these facilities as the network traffic grows. These are not the switch-to-cell site links but rather the switch-to-switch and/or switch-to-network node links responsible for call processing and database inquiries.

One prevalent type of interswitch and internode interconnection protocol used in the cellular industry is ANSI standard system signaling 7 (SS7). These types of data links are assumed for this report. Other such transmission protocol types will use similar metrics to measure data link performance. Thus, for evaluating link performance, this report should include, as a minimum, the following data fields:

Telephone Number Inventory Report

System Name: _____

Week Ending: _____

Date: _____

NPA	NXX	XXXX	Available	Central office	Resident switch	Route name	Tandem	Tested	# released	# active	Utilization rate	Comments
914	365	7000-7999	8/15/95	ORB	WDB1/003	BBEN9	ZERK	1000	1000	500	50%	*
201	968	2000-2999	8/26/95	PARM	ERU07/011	CPG2/4	NWK	1000	500	200	20%	

Figure 8.4 Sample telephone number inventory report.

Facilities Usage Report

System Name: _____

Month: _____

Date: _____

Trunk#	Usage	Band	% Band usage	% Variation previous month	Design %
NJ004-05	8545	1	25	4	25
		2	35	1	35
		3	30	5	25
		4	10	6	15

Figure 8.5 Sample facility usage report.

Basic link configuration data

1. Defined network point codes
2. Number of defined link sets in the network
3. Number of links defined in each link set

Performance data

1. Link traffic load (erlangs)—System busy hour data only
2. Link set traffic load (erlangs)—System busy hour data only
3. System traffic load (erlangs)—System busy hour data only
4. Link active and inactive times—Total peak service hours (0700–2000)
5. Link set change over count—Total peak service hours (0700–2000)
6. Link retransmission percent—Total peak service hours (0700–2000)

These data can be collected by using either the network switches or separate monitoring equipment such as a protocol analyzer patched into the network links in a nonintrusive (passive) manner. This report should be generated on a weekly basis with the responsibility of its review assigned to the data engineering group within the network engineering department. If, however, a problem is noticed on a link set or a link, a more detailed report about this problem facility should be obtained to conduct more in-depth troubleshooting. The actual

Facilities Interconnect Report

System name: System X

Date: _____

Network point code assignments:

Node name	255 - 1 - 1
Node name	255 - 1 - 2

Network link definitions:

Link set: 255 - 1 - 1	Links: SLC - 1
	SLC - 2
Link set: 255 - 1 - 2	Links: SLC - 1

Link traffic data:

Link set 255 - 1 - 1	0.38 erlang	SLC - 1 0.20 erlang
		SLC - 2 0.18 erlang
Link set 255 - 1 - 2	0.15 erlang	SLC - 1 0.15 erlang

System data link traffic load: 0.43 erlang
Link service data:

Link set 225 - 1 - 1	Links: SLC-1	No outages
	SLC - 2	0900–1000 hour,
		02:00 min
	Link changeover count:	2
	Link routing error count:	0
	Link retransmission %:	SLC - 1 0.15%
		SLC - 2 0.01%
Link set 225 - 1 - 2	Links: SLC - 1	No outages
	Link changeover count:	0
	Link routing error count:	0
	Link retransmission %:	SLC - 1 0.01%

Figure 8.6 Sample facilities interconnect report.

link/link set performance threshold settings are based upon the ANSI standards manual to ensure accuracy and uniformity. A sample facilities interconnect report format is shown in Fig. 8.6.

8.1.5 Switch/node metrics report

The switch/node metrics report is intended to assist in the monitoring and dimensioning of the switches in the network. This report should be generated once a week for each switch/node in the system and be provided to the network engineering department for review. It should include, as a minimum, the data fields in Table 8.1.

As mentioned above, this report should be generated once a week as part of the normal maintenance and operation of the network switches and nodes. However, if a problem is encountered or if any metric listed above approaches an operational limit, this report or the metric(s) of

TABLE 8.1 Switch/Node Metrics Report Data Fields

Performance data	Recommended sample times
Central processor (CPU) load and utilization value	System busy hour %
Secondary processors (SPs) load and utilization value	System busy hour %
Port capacity assigned	Weekly average
Port capacity available	Actual
Subscriber capacity assigned	Weekly average
Subscriber capacity available	Weekly average
Memory capacity assigned	Weekly average
Memory capacity available	Weekly average
Switch/node I/O capacity assigned	Weekly assignments
Switch/node I/O capacity available	Actual
Service circuits load and utilization values (for senders, receivers, tone generators, etc.)	System busy hour %
Switch/node outages (number and duration of occurrences)	Daily recordings

interest should be collected and reviewed on a daily basis. This will provide better accuracy when monitoring the switch or node of interest. The proposed switch/node metrics report is shown in Fig. 8.7.

8.1.6 System traffic forecast report

The system traffic forecast report is a plot of the total system usage monthly average for an entire year. The actual data are the 10 highest system busy hour days traffic for the network. This report and plot is used to predict yearly trends in the system traffic, which will show the season changes to the network traffic patterns. A yearly traffic trend report is shown in Fig. 8.8.

8.1.7 Network configuration report

The network configuration report is a collection of diagrams showing the network configuration as it exists. The network configuration report needs to be updated upon every major network change, i.e., new switch, node additions, new central office exchanges, etc. It should be distributed to all of network engineering and the operations department as well. The network configuration required tables and diagrams are:

1. Voice interconnect diagrams

   - Mobile system to PSTN LEC central offices and IXC tandems
   - Internal system voice facilities for call delivery and handoffs

---

Switch/Node Metrics Report

System name: System X

Date: _____

Report week: _____

Node: Switch 1
CPU Load and processor occupancy	47%
Secondary processor 1 load	29%
Secondary processor 2 load	20%
Secondary processor 3 load	23%

Node port capacity	985 / 1200 matrix ports
Node subscriber capacity	45,000 / 70,000 subscriber records
Node memory capacity	10 M / 35 M
Node I/O capacity	12 / 21 I/O ports

Service circuits loading:
Sender circuits	30%
Receiver circuits (MF / DTMF)	43%
Conference circuits	23%

Node outage data No outage for this report period
Start time of outage:
End time of outage:
Duration of outage:
Reason for outage:

Figure 8.7 Sample switch/node metrics report.

2. Data network diagram

   - SS7 data links to external networks
   - SS7 data links for internal system

3. System switch to cell site assignments

4. Auxiliary system interconnection diagrams

   - Voice mail
   - Validation systems

A sample of a network configuration report is shown in Fig. 8.9.

8.2 RF Performance Report

The RF performance report is meant to report to various engineering groups the current level of system performance the RF network is

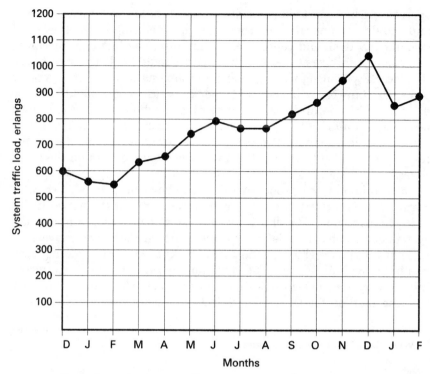

Figure 8.8 Example yearly system traffic plot.

Title: System X PSTN interconnection

Date: _____

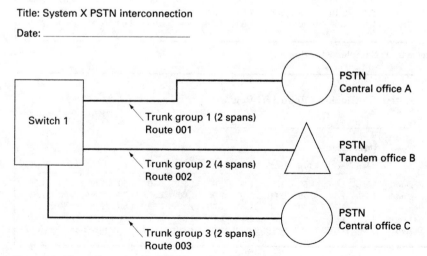

Figure 8.9 Example system to PSTN serving CO drawing.

operating. The report should contain some tabular and graphical data and should be generated on a weekly basis showing several levels of granularity. It should contain as a minimum the lost call, blocking, and attempt failure statistics. The report should be distributed to the RF and RF performance engineering groups as a whole on a weekly basis. On a biweekly basis the RF performance report should also be distributed to the director of engineering. It should be broken down into the granularity level that is a weekly average for the system. If the system is subdivided into individual areas of responsibility, the report should also have the individual regions displayed on the same chart.

An example of the proposed report is shown in Fig. 8.10 to 8.12. Figure 8.10 is a chart of the lost call performance of the network and individual regions. The chart for the lost call rate also contains the system design objective defined for the yearly goals and objectives.

Figure 8.11 is a chart of the radio blocking levels for the network and by individual regions. The chart for the radio blocking again contains the system design objectives. For this chart I have assumed a banding method for the design objective of the network. The design

System Lost Call Report by Region

Date:

Time: 1700

	1-Jan	1-Feb	1-Mar	1-Apr	1-May	1-Jun	1-Jul	1-Aug	1-Sep	1-Oct	1-Nov	1-Dec
Region 1	1.9	1.85	1.95	2.1	2.05	2.10	2.05	2.00	2.01	2.05	2.20	2.10
Region 2	2.1	2.10	2.05	1.90	2.00	1.98	2.00	1.87	2.03	2.01	2.00	1.99
Region 3	2.2	2.15	2.10	1.80	2.05	2.10	2.10	2.05	2.00	2.08	2.01	2.00
Network	2.1	2.10	2.05	2.00	2.05	2.05	2.07	1.98	2.00	2.03	2.10	2.00
Design	2.00	2.00	2.00	2.00	2.00	2.00	2.00	2.00	2.00	2.00	2.00	2.00

Figure 8.10 Sample lost call report.

System RF Blocking Report by Region

Date: _____

Time: 1700

	1-Jan	1-Feb	1-Mar	1-Apr	1-May	1-Jun	1-Jul	1-Aug	1-Sep	1-Oct	1-Nov	1-Dec
Region 1	1.9	1.60	1.50	1.4	1.30	1.80	1.70	1.60	1.50	1.60	1.70	1.70
Region 2	1.50	1.50	1.50	1.60	1.70	1.90	1.60	1.87	1.70	1.80	2.00	1.99
Region 3	1.7	1.70	1.60	1.80	1.90	1.50	1.40	1.60	1.80	1.30	1.80	1.40
Network	1.7	1.6	1.53	1.6	1.63	1.73	1.57	1.69	1.67	1.57	1.83	1.70
Design	1<>2	1<>2	1<>2	1<>2	1<>2	1<>2	1<>2	1<>2	1<>2	1<>2	1<>2	1<>2

Figure 8.11 Sample system RF blocking report by region.

System Attempt Failure Report by Region

Date: _____

Time: 1700

	1-Jan	1-Feb	1-Mar	1-Apr	1-May	1-Jun	1-Jul	1-Aug	1-Sep	1-Oct	1-Nov	1-Dec
Region 1	2.2	1.60	1.50	1.4	1.30	1.80	1.70	1.60	2.10	1.60	1.70	2.10
Region 2	1.50	1.50	1.80	1.60	2.10	1.90	1.60	1.87	1.70	1.80	2.00	1.99
Region 3	1.43	1.82	1.60	1.80	1.90	1.50	1.96	1.60	1.80	1.93	1.80	1.40
Network	1.71	1.64	1.63	1.6	1.77	1.73	1.75	1.69	1.87	1.78	1.83	1.83
Design	2.00	2.00	2.00	2.00	2.00	2.00	2.00	2.00	2.00	2.00	2.00	2.00

Figure 8.12 Sample system attempt failure report by region.

objective used on this chart should be the same as the yearly goals and objectives.

Figure 8.12 is a chart representing the attempt failure levels in the network and individual regions. The chart for the attempt failures has the system design objective as one of the lines displayed for reference. The design objective used in the chart should be the same that is used for the yearly goals and objectives. There is one other piece of information that needs to be generated on a regular basis with more frequency than that mentioned before. The information shown in Fig. 8.13 should be generated and distributed to the RF and performance engineers on a daily basis. Figure 8.13 is used to help define the performance of the network in a more refined level. It shows not only the lost call, RF radio blocking, and attempt failure levels but also usage and call completion ratios. The information presented can be used to help monitor the system for any peculiar items that occur on a daily basis.

8.3 System Growth Status Report

The system growth status report is meant to report on the various key elements associated with the system growth plan put out on a

System Report by Region

Date: _____

Region	Time	Usage	O&T	LC%	# LC	% AF	# AF	% Block	# Block	Usage/L
Region 1	1700	6,495	5,905	2.1	124	4.88	288	1.13	67	52.38
Region 2	1700	19,163	17,421	1.9	331	1.46	254	1.01	176	57.89
Region 3	1700	9,961	9,056	1.8	163	2.33	211	2.05	186	61.11
Network	1700	35,620	32,381	1.91	618	2.33	753	1.32	429	57.64
Design				2		2.00		1><2		>50

Figure 8.13 Same report by region.

quarterly basis. It is meant to ensure that the design requirements put forth in the growth plan are being met and still remain valid. This report is generated on a monthly basis and folded into the quarterly system growth studies that take place. The simple intention is to report on the status of the critical system growth indicators and make sure that the system is performing as expected for the growth predicted. The system growth status report is shown in Fig. 8.14 with sample numbers.

8.4   Exception Report

The exception report is one of the most invaluable tools available to troubleshooting on a daily basis. It should be generated by operations and engineering combined and should contain the basic information regarding the network from a maintenance point of view. The report should be distributed to the network manager, RF performance manager, and also the operations managers as a minimum. The directors for both engineering and operations would also benefit from seeing the report. The one primary problem that will occur is the coordination efforts. It might be best to have one group be directly responsible for the generation of the report and a key contact in the other group provide tactical data.

The operations report is meant to help identify the current level of maintenance issues in the network at any given time and to help improve the troubleshooting response for engineering. The information in this report is time-sensitive and its dissemination is critical to the parties that need it. It is strongly suggested that if an electronic mail system is used by the system operator this report be stored in a central location for quick reference. The exception reports structure is given in Fig. 8.15.

	Current	Design	Variance
Subscribers	250,000	255,000	5000
System BH erlangs	1000	1050	50
Radios	3126	3200	(74)
Cell sites	160	165	(5)
Erlang per subscriber	7	7.1	0.1
Switches	3	3	0
Lost call rate	2.1%	2.05%	2.4%
Attempt failure	1.8%	1.8%	0.0%
Blocking	1%><2%	1.4%	NA

Figure 8.14 System growth status report.

```
Exception Report

Date: 8/1/95
Previous day's busy hour data
    Total calls completed:              100,000
    Usage:                              75,000
    Lost calls:                         1.9%
    Attempt failures:                   1.85%
    RF blocking:                        1.5%
Cell site outage:                       1
T1 outage:                              2
Radios OOS:                             50
Planned outages (description)

Unplanned outages (description)
```

Figure 8.15 The exception report.

8.5 Customer Care Reports

The customer care reports are one of the key metrics for receiving information about the quality of the network. They should be integrated with the trouble reporting and resolution system that is used by customer care and performance engineering. The frequency of the customer care report should be biweekly and distributed to all the managers in engineering and operations. The level of detail and information content in the report is shown in Fig. 8.16.

```
Customer Care Report

System Name: _____

Date: _____
```

	Region 1	Region 2	Region 3	System
Network complaints	11	12	17	40
Lost calls	3	2	5	10
Interference complaints	5	9	5	19
Did not get onto system	3	1	7	11
Trouble reports issued	4	6	5	15
Trouble reports closed	3	6	5	14
Outstanding trouble reports	1	0	0	1

Figure 8.16 Sample customer care report.

8.6 Project Status Reports (Current and Pending)

The project status report is meant to track all the major projects currently underway or proposed by engineering for the coming year. It should have a rolling 1-year projection as the best possible position to report on. The report is meant to help identify major network activities that are taking place or will take place. It should match the organization and company's goals and objectives. This report needs to be generated and issued on a biweekly basis to all the managers for engineering, operations, and implementation. The objective is to ensure that all the technical community for the company is fully aware of all the projects engineering is currently or will be involved with. The format for the report is shown in Fig. 8.17.

It is imperative that when the project is formed the project plan include an executive summary, objective, project prime, review dates, milestones, labor loading, budget impact, impact to other departments, pass-fail criteria, and a MOP. When the project is completed, the postimplementation review and report needs to be generated by the project leader. The executive summary for the project plan and the postanalysis can easily be used for conveying the objective and final outcome of the project to upper management. The recommended general format is illustrated below.

Project plan report format

1.0 Executive summary (one page)
1.1 Objective
1.2 Expected time
1.3 Labor and infrastructure requirements
1.4 Projected cost
1.5 Positive network impact
1.6 Negative network impact

2.0 Project description
2.1 Project leader
2.2 Project team
2.3 Project milestones
2.4 Related documents

3.0 Design criteria
3.1 Basic design criteria
3.2 Project review dates
3.3 Method of procedure
3.4 Hardware changes
3.5 Software changes

Project Status Report

System Name: _____

Date: _____

Dept	Project name	Priority	Originator	Lead dept/person	Due date	Project plan	Capital funding	Comments
RF Eng	Alpha	1	Marketing	RF/Smith	8/15/95	7/16/95	7/17/95	On-target
Netwrk	Tree	2	Engineering	Netwk/Gervelis	8/15/95	7/20/95	7/24/95	On-target

Figure 8.17 Sample project status report.

3.6 Pass-fail criteria
3.7 Preimplementation test plan
3.8 Postimplementation test plan

4.0 Resources
4.1 Labor projections

- Weekly

- Monthly

4.2 Infrastructure requirements
4.3 Interdepartment resources required
4.4 External department impact

- Marketing

- Customer care

- MIS

- Operations

- Implementation

- Vendor

5.0 Budget
5.1 Total project cost
5.2 Capital budget impact
5.3 Expense budget impact
5.4 Comparison of project costs to budget, planned and actual

8.7 System Status Bulletin Board

The system status bulletin board is an essential element in the continued process of conveying critical system information in a uniform and public format. It is a central location where network data regarding system performance should be exhibited. The bulletin board should be displayed on a centralized wall in the engineering department to enable other company personnel, besides engineers, to view the data and possibly comment on it. The bulletin board, if used correctly, will foster improved motivation by demonstrating the group's hard work. The material that should be included in the system status board is listed in Fig. 6.7.

8.8 System Software Report

The system software report is another essential report (Fig. 8.18). Its purpose is to identify to all the managers and relevant engineers in

System Software Report

System: X

Date: _____

	Current	Tested	Next load and expected date
Switch CPU			
Switch matrix			
Switch database			
Voice mail DACs			
Cell site			

Figure 8.18   System software report.

engineering and operations the current software configuration for the network. The report should be only one page in length and issued on a biweekly or monthly basis.

8.9   Upper Management Reports

The upper management reports are an essential element in RF and network operation. The use of a series of quick and concise information to upper management will palliate their thirst for critical knowledge of the performance of the network. The amount and types of reports that can be sent to the upper management of any company range from voluminous to sparse. It is recommended that the information in Fig. 8.19 be produced and disseminated to all the management in the technical area of the company on a monthly basis. The information to be issued upward contains data extracted from most of the reports recommended in this chapter. The difference in the reports lies mainly in the format and amount of information content delivered.

The proposed upper management report is shown in Fig. 8.19.

8.10   Company Meetings

Meetings are an essential form of report generation. The type and frequency of the actual meetings are indicative of the focus placed on designing and improving the network. The critical point is that there

---

Management Report

System Name: _____

Date: _____

	Present	Prior month	Variance

System Measurements
 Total Mobiles
 Calls Handled
 Total System Erlangs
 Avg BBH Erlangs
 Erlang/Sub
RF
 % Lost Calls
 % RF Blocks
 % Attempt Failures
 % Radios OOS
Network
 CPU Load/Utilization Rates
 Subscriber Capacity
 Switch/Node I/I Capacity
 Memory Capacity
Configuration
 Switches
 Cell Sites
 WPBX
 RF Radios
 Microwave T1's
 PSTN Trunks
 Total Trunks
 DACs
 POPs

Figure 8.19   Upper management report.

needs to be an effective balance between the number of meetings held
and the need to convey information in a person-to-person environment. It is recommended that you review the current meeting levels
you are currently using and see if you are meeting either too frequently or not enough. A suggested meeting structure is proposed for
various levels in the organization:

Director to director	Once a week
Director to manager	Once a week
Manager to manager	Once a week (you need to talk with your counterparts on a more fre-quent basis)

Engineering and marketing	Once a month
Engineering department meetings	Once a month
Engineering and customer care	Once a month
Engineering design reviews	Once a week
Project approval meeting	Biweekly
Network growth plan	Every 3 months

Except for the engineering design reviews, project approval, and network growth plan the meetings should not last more than 1 hour. It is recommended that the weekly status meeting between the groups take place at the beginning of the week instead of the end of the week. An agenda for each of the meetings should also be provided prior to the meeting so it stays focused.

Regarding company meetings, it is imperative that the marketing department plan initial meetings with the engineering department prior to the onset of any new product or service to assess the feasibility of the offering. The meetings between marketing and engineering can take place more frequently than once a month. The main and auxiliary system capabilities should be discussed with marketing as well to ensure that these system capacities can support the projected growth. The marketing departments need to provide for any new product and services they plan to launch to the customer base.

An example of poor coordination occurred when marketing began selling voice mail as part of a new service offering without consulting with network engineering. Many customers bought this service but the auxiliary voice mail system could not support their rapid growth. Customers became angry when they realized they purchased a product but once doing so were told it would not be available for another 6 months.

It is important that the following key elements, or rules, be followed to ensure company meetings are successful. A meeting's success is defined as having met its stated purpose with a decision being reached or the required follow-up action items being defined and delegated.

Premeeting steps:

1. Plan what the meeting is about and its desired outcome.

2. Identify who should attend the meeting.

3. Develop the meeting agenda and distribute it in advance of the meeting.

4. Ensure that there is sufficient room and audio-video equipment for the meeting.

Meeting steps:

1. Introduce everyone and clarify everyone's role in the meeting.
2. Review the agenda.
3. Cover each agenda item one at a time.
4. Allow for sufficient feedback on each topic and control defocused people.
5. Close all discussions on the topic at hand before moving to the next agenda item.
6. Summarize all decisions and agree upon action items.
7. Draft agenda for next meeting and agree upon a time.
8. Close meeting by thanking everyone for attending.

Postmeeting steps:

1. Write and distribute meeting minutes promptly.
2. File agenda, minutes, and other key documents in engineering library and LAN.
3. Follow up on all open items to ensure closure.

8.11   Network Briefings

Network briefings are important for reporting the status of the network, or state of the union, to the rest of the company on a regular basis. Specifically the network briefing is a once a month open dialog between the technical departments and the remaining company departments. The network briefing enables a large volume of information to be conveyed and a uniform dissemination of information to take place.

It is recommended that the network briefings take on a format that is uniform and also timely for the general audience that will attend. The network briefings should take place on a monthly basis at a regular predetermined time. The meeting should be scheduled to last no longer than 2 hours. The choice of an overhead presentation or a more informal approach is dependent upon your company's culture. However, the format for the network briefing is as follows:

1. *Introduction.*   Introduction of the various speakers and the agenda to be used for the meeting.

2. *Engineering Activities.*   This is where the last month's activities are talked about, usually a follow-up to the previous month's comments. In addition the current month's activities are discussed with

expected outcomes. The general nature of the talk is high-level and not detail-oriented.

3. *Construction.* This part of the talk involves a representative from implementation and real estate providing a talk regarding the current build program for the company. The topics to cover involve the recent completion of sites for the network. In addition to the sites built, the sites currently under construction are also discussed.

4. *Performance.* A member of the RF performance group presents a brief presentation regarding the key performance metric for the network. The metrics usually should be lost calls, blocking, and attempt failures. The metric charts used should be the same charts displayed in engineering for consistency.

5. *Planned network and technical projects.* The facilitator of the meeting then goes over the projects and major events that are planned to take place between this meeting and the next with respect to the network.

6. *Discussion.* This is the section where members of the audience ask questions regarding any network issues they have.

7. *Closing.* The meeting is closed and the next meeting date and location are mentioned.

8.12 Report Flow

The reports that need to be generated in the technical organization are vast in number if all the possible issues are pursued. However, there is never enough time and worker resources available to pursue all the technical issues for a network that is constantly growing. Therefore, the primary focus needs to be placed on what reports should be utilized on a regular basis to ensure the health and well-being of the network. Table 8.2 is a brief breakdown of the various reports that need to be generated and distributed within the technical organization. The focus of the reports generated and received pertains to the engineering department as a whole. The organization utilized for the report structure listed here is based on the organization structure of Fig. 8.20. Chapter 10 discusses organization recommendations.

References

1. AT&T Bell Laboratories, *Engineering and Operations in the Bell System,* Murray Hill, N.J., 1984.
2. Ericsson Radio, Inc., Signalling System Number 7 Theory and Operation—Course Number 140.
3. Lee, William C. Y., *Mobile Cellular Telecommunications Systems,* McGraw-Hill, New York, 1989.
4. Motorola, Inc., *System Description Manual,* Arlington Heights, Ill., 1987.

TABLE 8.2   Reports Needed within the Technical Organization

Department/Report	Internal/External	Frequency
Director engineering		
Key performance	External	Monthly
Facilities	Internal	Monthly
Growth plan	External	Quarterly
Software loads	Internal	Monthly
Project plan	Internal	Weekly
Cell site progress		
System performance	Internal	Weekly
Equipment status	Internal	Weekly
Network briefing	External	Monthly
Directory inventory	External	Monthly
Network systems		
Individual node performance report	Internal	Weekly
Network link performance report	Internal	Weekly
Network routing report	Internal	Weekly
Network software report	Internal	Biweekly
Facilities management		
Systems facility interconnect plan	Internal	Quarterly
Data network		
Systems facility interconnect plan	Internal	Quarterly
Network link performance report	Internal	Weekly
Software engineering		
Software configuration	Internal	Weekly
FOA	External	Biweekly
Translations status	Internal	Weekly
RF engineering		
Cell site status	External	Weekly
FAA compliance	External	Weekly
Frequency management	Internal	Weekly
RF design	Internal	Weekly
Digital radio (CDMA/TDMA)	Internal	Weekly
CDPD	Internal	Weekly
System performance		
Performance report	External	Weekly
Equipment engineering		
Equipment status	External	Weekly

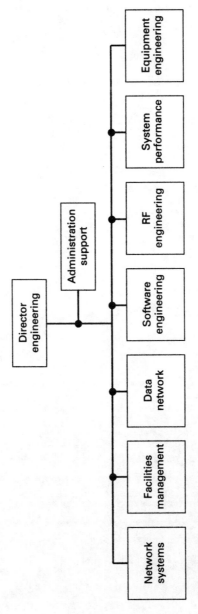

Figure 8.20 Organization.

9

Network and
RF Growth Planning

The network and RF growth planning for any network is exceptionally critical because it defines the how, what, when, and why aspects for the company's technical direction. The direction put forth from this plan is used to define most, if not all, the projects that need to be completed over the report's time frame by the technical community in the company.

This chapter discusses how to actually put together a growth plan which incorporates the network and RF portions in the design process. The network and RF design process are linked in various points in the growth planning, and it is necessary to incorporate both aspects to ensure a proper design is completed.

9.1 Methodology

The methodology for the network and RF growth plan needs to be established at the beginning of the process to ensure the proper baseline assumptions are used and to prevent labor-intensive reworks. Some of the issues that need to be identified at the beginning of study:

1. Time frames for the report

2. Subscriber growth projections

3. Design criteria

4. Baseline system numbers for building on the growth study

5. Cell site construction expectations

6. New technology deployment

7. Budget constraints

8. Due date for report

9. Maximum and minimum offloading for cell sites

It is essential to establish the time frames for the report prior to the report generation. They will define what the baseline, foundation, and future look of the report will present. The baseline month or time frame used for generating the report is critical, since the wrong baseline dates will alter the outcome of the report.

The amount of time the report projection is to take into account is also critical for the analysis. The decision to project 1 year, 2 years, 5 years, or even 10 years has a dramatic effect on the final outcome. In addition to the projection time frame it is important to establish the granularity of the reporting period, monthly, quarterly, 6-month, yearly, or some perturbation of them all.

The particular marketing plans also need to be factored into the report itself. The marketing department plans are the leading element in any network and RF growth study. The basic input parameters to the network and RF growth plan provided by the marketing department are listed next.

1. The projected subscriber growth for the system over the time frame for the study

2. The projected millierlangs per subscriber expected at discrete time intervals for the study

3. The dilution rates for the subscriber usage over the time frame for the study

4. Types of subscriber equipment used in the network and percent distribution of CPE projections (i.e., portable or mobile units in use and their percent distribution)

5. Special promotion plans over the time frame of the study like local calling or free weekend use

6. The projected number of mobile data users over the time frame of the study

7. Top 10 customer complaint areas in the network requiring coverage improvements

A multitude of other items are needed from the marketing department for determining network and RF growth. However, if you obtain the information on the basic seven topics listed above from the marketing department it will be enough to start the network and RF growth study. The establishment of the design criteria used for the report is another key criterion when putting together the plan. It is recommended that the design criteria used for the study be signed off by the director of engineering to ensure that nothing is missed.

The items in the RF and network growth plan design criteria should include the following as a minimum.

1. Marketing input

2. RF spectrum available for the growth plan

3. Type of grade of service table to be used for plan, i.e., erlang B P02

4. Minimum and maximum offloading factors for new cells

5. Coverage requirements

6. Identification of coverage sites

7. BH peak traffic, 10-day high average for month

8. Infrastructure equipment constraints

9. Digital and analog radio growth

10. New technology considerations

11. Cell site configurations used for new cells

12. Baseline system numbers

The above list should be used as the foundation for establishing the design criteria for the RF and network growth plan.

The baseline system numbers should be listed in the design criteria for the growth plan. The baseline numbers involve defining the time period for basing the growth projection. In particular the decision to, say, use June data instead of July data has a significant impact in the plan as well as using peak versus an average value for traffic.

The construction aspects of the system will also need to be factored into the report itself. They pertain to the proposed new cell sites, or network equipment, that are or will shortly be under construction. Other construction aspects involve the possibility of actually building what is requested in the specified time frame dictated in the report. The realistic time frame for construction is the reality dose that needs to be placed in the report saying that although 100 sites are needed in the next 3 months only 20 will actually be available for operation.

New technology deployments also need to be factored into the plan. They might be converting from a TDMA system to a CDMA system. Other technology deployment issues that need to be factored in might be the cell site infrastructure equipment, where a new vendor's equipment will be deployed. Also involved with technology deployments could be the introduction of a new switch that has improved processor capacity for key trigger points which change the requirements for future growth.

The budget constraints imposed on any operator must be folded into the network and RF growth plan. Failure to incorporate them into the plan will only make the plan unrealistic and squander everyone's time needlessly. When the report is put together any budgetary constraints that exist must be incorporated into the plan. Often if the plan calls for more sites than are budgeted for, as is the case most times, it is imperative to help establish a ranking order for which sites are truly needed versus which are nice to have. The last part of the preparation for the report is finding out when the report itself is due, the level of detail required, and who is assigned to work on the report itself.

9.2 Tables

Several tables should be used for putting together the network and RF growth plan. Table 9.1 is an erlang B table. With the erlang B equation (9.1) you can construct a traffic look-up table in your network and RF growth plan.

TABLE 9.1   Erlang B Table

| Channels | Blocking probability | | | | | |
	P01	P012	P015	P02	P03	P05
1	0.0101	0.0121	0.0152	0.0204	0.0309	0.0525
2	0.153	0.168	0.19	0.223	0.282	0.381
3	0.455	0.489	0.535	0.602	0.715	0.899
4	0.869	0.922	0.992	1.09	1.26	1.52
5	1.36	1.43	1.52	1.66	1.88	2.22
6	1.91	2	2.11	2.28	2.54	2.96
7	2.5	2.6	2.74	2.94	3.25	3.74
8	3.13	3.25	3.4	3.63	3.99	4.54
9	3.78	3.92	4.09	4.34	4.75	5.37
10	4.46	4.61	4.81	5.08	5.53	6.22
11	5.16	5.32	5.54	5.84	6.33	7.08
12	5.88	6.05	6.29	6.61	7.14	7.95
13	6.61	6.8	7.05	7.4	7.97	8.83
14	7.35	7.56	7.82	8.2	8.8	9.73
15	8.11	8.33	8.61	9.01	9.65	10.6
16	8.88	9.11	9.41	9.83	10.5	11.5
17	9.65	9.89	10.2	10.7	11.4	12.5
18	10.4	10.7	11	11.5	12.2	13.4
19	11.2	11.5	11.8	12.3	13.1	14.3
20	12	12.3	12.7	13.2	14	15.2

$$\text{Grade of service} = \frac{E^N/N!}{\sum\limits_{n=0}^{N} E^n/n!} \qquad (9.1)$$

where E = erlang traffic and N = number of trunks (voice channels) in the group.

9.3 Switch Growth

The switch growth study will be broken down into three major areas of interest. The first critical metric is the processor capacity study, the second important metric is the port capacity study, and the final metric to review is the mobile subscriber database capacity study.

9.3.1 Switch processor capacity study

Many methods and models are used to determine the current processor capacity of a switch or node in a particular network and to predict when an upper threshold will be reached due to increased traffic loads. In this section we discuss only a simple study for use in managing your network switches and/or nodes. More complex and accurate studies can be formulated by working with your equipment vendors and obtaining their assistance in measuring your unique system loads.

A typical processor load study consists of the following steps. First, assemble the design assumptions for the study. Then begin measuring the actual processor load(s) of the switch. Once the data are available, plot the processor load values on a graph and then project this plot into the future for a specified period of time (i.e., 6 months, 1 year, etc.). This can be accomplished in one of two ways. By using regression analysis on the data collected the best-fit straight-line approximation for the data can be obtained. The plot of this line would represent the projected load of the processor for the coming months in the system, assuming uniform traffic growth. However, for a more accurate projection use the system subscriber forecast developed by the marketing department. (More on this method later.) The next step is to obtain the acceptable maximum threshold operating limit of the processor from your switch vendor (with acceptable variance) and mark this limit on the plot.

The final step in the processor capacity study involves the analysis of the data and the planning of the network for relief of the processor overload. The point where the processor load plot intersects the maximum limit threshold mark is the time period in which planned relief of this switch should take place by either expanding the network (adding more switching capacity) or performing system parameter changes (reduce registration intervals, etc.).

Steps for conducting a switch processor load study

1. Assemble and list all design assumptions of the study.

2. Measure and collect all processor load data; then plot these data vs. time on a graph.

3. Project this plot using either regression analysis or actual subscriber forecast data from the marketing department.

4. Obtain switch vendors' suggested processor load limit with an agreed-upon variance and plot this on the processor load graph. Note the time when the load of the processor reaches the specified upper threshold.

5. Present all data and analysis to your upper management during the quarterly design review. Develop a plan for expanding the network or for adjusting the proper switch and system parameters to relieve this processor load in the time frame provided by the study.

Example of a switch processor capacity study:

1. Assumptions for the system X processor capacity study:

- The actual measured data from the switch were collected during the system busy hour for the 10 highest traffic days of the month for a period of 6 months (October 1994–March 1995).

- The processor load was projected to the end of the year using both the regression analysis method and the subscriber forecast developed by the marketing department. Using the regression analysis method the traffic growth of the system is assumed constant and is based upon the previous 6 months processor load data. Using the data supplied by the marketing department the processor load can be related to the expected subscriber growth patterns in the following manner. Take the current traffic load on the system at 1000 erlangs. Then take the current total number of subscribers in the system at 200,000. Find the expected amount of traffic each subscriber contributes to your network by dividing the total system traffic (1000 erlangs) by the total number of subscribers in your network (200,000). This value equals 0.005 erlang per subscriber. It would be fair to assume that subscribers added to the system in the future would have similar calling patterns for your region and thus would contribute about the same amount of traffic to the network.

Now take the subscriber growth data from the yearly marketing forecast. Add the number of subscribers expected to be added to the system every month to the current total subscriber count. Then multiply this value by 0.005 erlang per subscriber. This gives you the traffic increase on the system on a monthly basis. By assuming the system traffic to be directly proportional to the processor load you can now obtain the projected processor load on a monthly basis as well. See Table 9.2 and Fig. 9.1 for more details.

- The vendor recommended maximum processor load is 75 percent with a 25 percent overload factor.

- The baseline load (no call traffic present) for the switch processor as specified by the vendor is 15 percent.

2. Measure and collect all processor load data and plot the results on a graph.

TABLE 9.2 System Data for Market X for Processor Load Study, March 1995

Current subscriber count	200,000
Current system load (busy hour erlangs)	1000
Current erlangs per subscriber	0.005
Current processor load (monthly average)	60%

Marketing subscriber growth forecast for the 1995 year

Assume a 3 percent subscriber growth rate for the months of April to September. Then increase the rate to 8 percent owing to the expected sales from the holiday season marketing promotions.

Month	Sys. Subs.	Month	Sys. Subs.	Month	Sys. Subs.	Month	Sys. Subs.
Jan.	188,000	Apr.	206,000	Jul.	224,000	Oct.	252,000
Feb.	194,000	May	212,000	Aug.	230,000	Nov.	268,000
Mar.	200,000	Jun.	218,000	Sep.	236,000	Dec.	284,000

Projected system traffic levels (erlangs) and proportional processor loads (%)

Month	Sys. Tr. (erl)	Proc. Load (%)	Month	Sys. Tr. (erl)	Proc. Load (%)
Apr.	1030	61.8	Sep.	1180	73.8
May	1060	63.6	Oct.	1260	78.6
Jun.	1090	65.4	Nov.	1340	83.4
Jul.	1120	67.2	Dec.	1420	88.2
Aug.	1150	69.0			

Actual measured processor load data

Year	Month	Proc. Load (%)	Year	Month	Proc. Load (%)
1994	Oct.	56.2	1995	Jan.	56.4
1994	Nov.	58.8	1995	Feb.	57.5
1994	Dec.	59.0	1995	Mar.	60.0

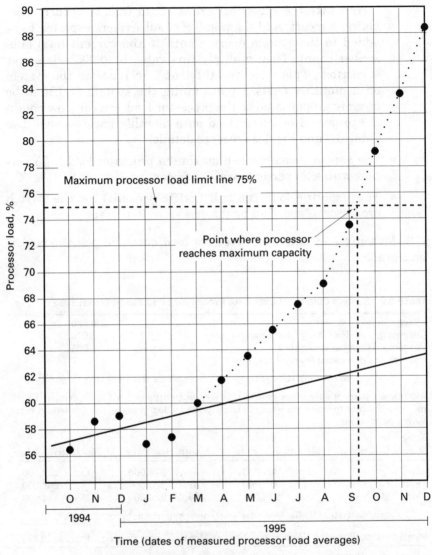

Figure 9.1 Switch processor load plot for market X.

3. Project the processor load plot.

4. Specify the maximum processor load limit and mark this load level on the processor load plot.

5. Report the time period when the switch will reach its projected maximum processor limit. According to the plot in Fig. 9.1 the switch for system X will reach a maximum processor load in early October 1995.

9.3.2 Switch port capacity study

The switch port capacity study is meant to take the current number of voice circuits assigned in the switch matrix and the total number of projected ports required over a given time period and determine if a maximum capacity will be exceeded. Every switch has a limited number of ports available for assignment. This value(s) can be obtained from your switch vendor and used to conduct your own port capacity study. The actual ports assigned to a switch can be broken down into two different types, RF voice circuits and standard landline interface circuits. Each type of circuit requires a different piece of switch hardware or magazine for interfacing to either a cell site or a landline facility. The maximum number of actual switch ports available for assignment will differ for both types, so make certain to study each for possible capacity limitations.

Once the switch port capacity limits are obtained, the next step is to determine the current port assignments for the switch. These data can be collected by printing out the database of the switch pertaining to the actual assignments of both landline and RF circuits. Tabulate the total number of both port types and begin the next step.

The next step will require input from the RF and network engineering groups. For the RF engineering group the projected number of coverage cell sites to be added to the system over the specified time period is required. For the network engineering group the projected capacity cell sites and the number of landline circuits to be added over the same time interval is required. These inputs determine the projected port growth for the switch. The actual data should be tabulated in such a manner that the projected port growth is shown in month-by-month increments. See Table 9.3 for an example of a system data report and Fig. 9.2 for an example of a port capacity plot.

Projected switch port capacity expansion. For the remainder of the 1995 year engineering expects to add 15 coverage cell sites and 8 capacity cell sites. The RF and network engineering groups estimated that one T-1 span (24 circuits) would be all that was required per each cell site. In addition to these ports the network group estimated that a total of 240 landline ports would be added to the system during the

TABLE 9.3   **System Data for Market X for Switch Port Study, March 1995**

Current number of RF switch ports available	1500
Current number of switch ports (RF) assigned	900
Maximum switch port capacity (RF)	2000
Current number of land switch ports available	2000
Current number of switch ports (land) assigned	1500
Maximum switch port capacity (land)	10,000

Month	RF Ports / Channels Additional / Total*	Land Ports / Circuits Additional / Total†
Apr.	3 cells = 72 ports / 972	48 / 1548
May	4 cells = 96 " / 1068	72 / 1620
Jun.	3 cells = 72 " / 1140	48 / 1668
Jul.	3 cells = 72 " / 1212	24 / 1692
Aug.	4 cells = 96 " / 1308	— / 1692
Sep.	2 cells = 48 " / 1356	48 / 1740
Oct.	2 cells = 48 " / 1404	— / 1740
Nov.	2 cells = 48 " / 1452	— / 1740
Dec.	/ 1452	— / 1740

*Total RF ports added: 552.
†Total land ports added: 240

remainder of the 1995 year. These data are tabulated below on a month-by-month basis.

A review of the system data and the switch port capacity plot indicates that if the current trend in RF port additions continues a new switch will be required to provide additional capacity some 8 months past November 1995, around July 1996.

Steps for conducting a switch port capacity study

1. Obtain the switch port capacity limits from your vendor.

2. Collect the projected coverage cell sites, capacity cell sites, and landline ports to be added in the time specified by the study from the engineering department.

3. Plot the RF and landline port growth data and predict when a maximum limit will be exceeded. This is the point in which additional port capacity will be needed in the network.

9.3.3   Switch subscriber capacity study

The switch subscriber growth study will mainly apply to smaller systems since networks of larger magnitudes have begun to implement HLRs (home location registers) as a central database storing their subscriber records for the system. However, this basic study can be applied to HLRs as well. As in the port capacity study discussed above the switch has a limitation on the number of subscriber records

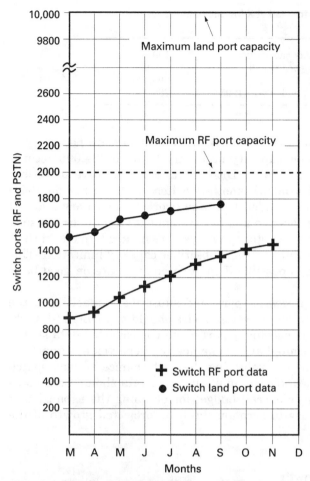

Figure 9.2   Switch port capacity plot for system X.

that can be stored at a given time. Again, this value will be available from your switching vendor along with any assistance in conducting this and other studies for managing your network.

Once the switch subscriber capacity limit is known, the following data are required. A copy of the subscriber forecast from the marketing department, as mentioned in the previous studies, will give the subscriber growth on a monthly basis. Also required are any special case numbers for use in network call delivery such as dynamic routing numbers (DRNs). (DRNs are numbers assigned in the switch for completing calls to your subscribers roaming in other systems.) Assemble these data in tabular form and analyze them for determining the subscriber capacity needs of the system. See Table 9.4.

TABLE 9.4   System Data for Market X for Subscriber Capacity Study, March 1995

Maximum switch subscriber capacity	100,000
Number of assigned DRNs per switch	300
Current subscriber count	200,000
Projected subscriber count end of year 1995	284,000

Based upon the data in Table 9.4, three switches are necessary to provide the subscriber capacity necessary to store 284,000 records plus the minimal, but additive, DRNs. With a conservative subscriber growth of 3 percent a month another switch or subscriber database node would have to be added to the network by February 1996. If your network is interfaced to another system with a set of established data links, it is possible to store your system's numbers in one of their nodes for temporary relief of this subscriber capacity limitation. This sort of arrangement is possible if the two markets are on good business terms.

In summary the processor, port, and subscriber capacities of the network switches and other system nodes should be monitored weekly by the network engineering department. The collected data should be analyzed and presented at the engineering department's quarterly design review for use in evaluating the performance of each switch and the system as a whole. In addition, these data should be used as input to the company's overall budget for planning the expansion of the network necessary to accommodate the projected growth of the system.

9.4   RF System Growth

The cell site growth projects in a network are an ongoing process of refinements and adjustments based on a multitude of variables, most of which are not under the control of the engineering department. However, the cell site growth analysis can be used to help direct the limited resources of the company. The final output of the cell site growth section of the plan is to identify the number of cell sites required for the network and their required on-air dates.

The process suggested for putting together the cell site and radio growth requirements as discussed here assumes that no automated cell building program is in place for your network. Such a program involves decision support systems which distribute traffic loads according to cell site parameters input into the modeling system.

9.4.1 Coverage requirements

The first step in the RF system growth process is to determine all the coverage requirements needed for the network. The total coverage requirements needed for the network will be rank-ordered in the final process to ensure that all the proper input parameters are taken into account.

RF coverage identification process

1. Coverage requirements identified by

    Marketing and sales

    System performance

    Operations

    Customer care

    RF engineering

2. Generate a propagation plot of the system or subregions that reflects the current system configuration.

3. Generate a propagation plot of the system or subregions that reflects the current and known future cell sites that are under construction.

4. Utilizing physical field measurements, generate a plot of the system or subregions.

5. Using plots from parts 2 and 3, compare this against areas identified in part 1 for correlation.

6. Using the field measurement plots, compare it against parts 1 and 2 for correlation.

7. Using the design criteria used by RF engineering, determine how many sites will be needed to satisfy the design goals.

8. Using the list of cell sites identified in step 7, rank-order them according to the following point system methodology.

The point system will involve five key parameters all ranked from a scale of 1 to 5 based on their severity. Each of the five categories receives a value which is then multiplied by each field to arrive at a ranking. The ranking methodology uses the following key fields:

1. Coverage

2. Erlang potential

3. Customer care problems

4. Marketing and sales needs

5. System performance requirements

To help clarify the ranking methodology, Table 9.5 with values added will assist in the explanation. It shows that cell 102 has a higher weight than 101 in the identification of prioritization. The ranking methodology should be applied to all the potential cell sites in a network. The rationale behind establishing a ranking system folds into the budgetary issues of having a method in place to determine where to cut or add sites to the build program.

9.4.2 Capacity cell sites required

The next step in the RF system growth plan is to determine the sites required for capacity relief. They involve a process of determining which sites and/or sectors expend their current capacity and require some level of relief, which can come from a variety of options listed below:

1. Radio additions

2. Parameter adjustments

3. Antenna system alteration

4. New cells

More capacity relief methods are available for redistributing the traffic loads of cell sites.

The first step in identifying the capacity cell sites required for a network involves use of a spreadsheet to determine where the problems are anticipated. The recommended spreadsheet format is shown in Table 9.6 with supporting equations following. The format shown is then carried out for each of the quarters involved in the initial study.

Obviously when looking at Table 9.6 a recursive feedback loop process is really implemented. The recursive feedback loop involves additional offloading and acquired erlang values until the proper design criteria are met.

TABLE 9.5

Cell	Coverage	Erlang potential	Customer care	Marketing	System performance	Total
101	3	1	2	4	2	48
102	2	4	3	1	4	96

TABLE 9.6   RF Growth Plan

System X

Date: ———————————————

RF growth plan

Quarter of interest (4Q95)

Cell	Baseline erlangs	channels	Channels	Offload	Acquired	Adjusted channels
1A	5.4	17	21	(0.4)	0	15
2A	4.0	12	16	0.0	0	16
101	0	0	0	0.6	3.1	12
XXX	a	b	c	d	e	f

[a] This is the baseline erlang value used to begin the chart. The baseline erlang value is arrived at based on the data collected in the design criteria. For example, if the baseline is for the average of the 10 busiest days in June the erlang value for this site's contribution is then deposited here.

[b] This is the number of physical channels currently available at the site during the baseline.

[c] This value is the number of channels needed for the site based on traffic projections arrived at in the design criteria. The channels needed are pulled from a look-up table, usually erlang B, which cross-referenced capacity to a physical number of channels. An example of the erlang B chart is in Table 9.1. The value that is used for the look-up is arrived at using this equation:

System erlangs projected
    = (millierlang/subscriber)*(total system subscribers for quarter)

Cell site system capacity* = (erlangs for sector)/(total system erlangs)

The part marked with an asterisk uses the data from the previous quarter. The value will not be 100 percent correct but is more than sufficient to use for planning purposes.

Cell site projected erlangs = (system erlangs projected) (cell site system capacity)

The cell site projected erlang value is then used in the look-up table for determining how many radios are needed for the site during this quarter.

[d] The offload value is used for shedding traffic to another cell site. The methodology used for arriving at how to shed traffic is discussed later. However, if it is determined that you can shed 40 percent of the site's projected traffic to another cell site, a value of 0.4 is entered.

[e] This is the acquired traffic the site receives from adjacent cell sites as a result of adjustments made to them. The acquired traffic portion should equal the total amount of traffic offloaded in section d. The values arrived at in this section pertain to how the traffic is distributed in the offloading cell site. For example, cell 5 may offload cell 3 by 40 percent, 0.4, but the traffic will not all go onto one sector, usually. Therefore, you must determine how this is distributed around the site. Continuing on this example, cell 5 will have cell 3's traffic distributed, with sector 1 getting 70 percent and sector 2 getting 30 percent. I strongly suggest keeping meticulous records about how much each site offloads the other and the distribution percentages used.

[f] The adjusted channels is the resultant of offloading and acquired traffic to the site. The formula to use in this example is

Adjusted erlangs = (cell site projected erlangs)*offload value + acquired traffic

The adjusted erlang value is then used with the same erlang look-up table used for c, and the RF channels needed are then listed.

Regarding the acquired traffic for a new site it is suggested that a minimum value be added to the overall value for every sector added. The added erlang value will be used to ensure that a minimum of one radio channel is put in place for every new cell. However, doing this will inflate the overall system erlang value beyond the projection for that quarter only.

9.4.3 RF traffic offloading

The offloading of traffic to another cell site can be accomplished through an elaborate computer simulation method or educated engineering guesses. The elaborate computer modeling capability is the ultimate desired method since it eliminates some guesswork involved in arriving at the offloading values. However, the computer model method is only as good as the model itself. If you are using the computer model method the offloading and acquired values should be readily extracted from the algorithm output and input into the spreadsheet either manually or automated.

If you are not fortunate to have a sophisticated computer modeling method, the following method is recommended. It has been used repeatedly with a high level of success. Referring to Fig. 9.3, a four-cell subsystem is listed for ease of illustration. The situation for this case will not involve the need for the site involving coverage-related issues. The example here involves only the introduction of a new cell site for the sole purpose of providing traffic relief to adjacent cell sites.

Traffic projections were done using the spreadsheet method listed above and it was found that sites 3 and 4 required some form of traf-

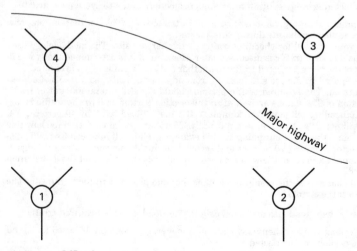

Figure 9.3 Offload.

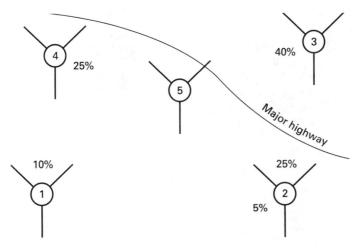

Figure 9.4 Offload cell.

fic shedding. It was determined that the only available method to ensure sufficient traffic shedding for both sites was the introduction of cell site 5 shown in Fig. 9.4. Referring to Fig. 9.4, cell site 5's placement in the network can ensure that it will offload several of the adjacent cell sites by the percentages listed. The offloading values were determined through an interactive process involving the following data:

1. Current coverage of zone by the existing cell sites
2. Determination of voice and control channel dominant server areas, by cell and sector
3. Evaluation of the cell site configurations in the zone
4. Incorporation of cell 5 into data provided for parts 1, 2, and 3
5. Comparison of propagation plots to actual field measurement data
6. Establishment of offloading and acquired percentages

The offloaded traffic is then assigned to the sectors of cell 5 as shown in Fig. 9.5. The values represented in Figs. 9.4 and 9.5 are then inputted into the tables of the growth plan to ensure that the desired effect is accomplished.

9.5 Radio Growth

The radio growth portion of the network and RF growth plan involves analysis of the individual sectors for every cell site currently in the network and proposed. The radio growth projections are then com-

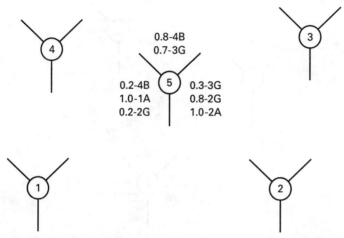

Figure 9.5 Offload percentages.

piled and fed into the traffic performance process discussed in Chap.
6. The radio growth of the network has a large impact on the capital
and recurring costs.

9.6 Network Interconnect (Voice)

The network interconnect growth study (voice) is the collection of current transmission facilities [T-spans, DS-3's, POTS lines, microwave
links, etc., either purchased from the LEC (local exchange carrier) or
privately owned by the company itself] and the assessment or analysis as to whether these facilities need to be expanded or decommissioned. These facilities are the required medium to interconnect the
mobile switches to the PSTN (public switched telephone network) and
each individual cell site for providing cellular service to area mobile
subscribers. For ease of accounting and management the term facilities mentioned above should include only the voice circuits if possible.
The facilities used for the network data links should be broken out
into another study. See Sec. 9.7 for more details.

To conduct this study first collect the current system interconnect
data. These data include all the voice circuits or trunk groups interconnecting the network switches. Note, an individual voice circuit is
sometimes referred to as a trunk. Thus a group of trunks is called a
trunk group. Trunk groups that interconnect network switches are
sometimes referred to as IMTs (intermachine trunks). See Chap. 3 for
more details. Also considered as interconnect data are the trunk
groups to the PSTN. These include trunks to the LEC and the IXC
(interexchange carrier) as well. The next major category of facilities

are the cell site trunks. The number of trunks assigned to an individual cell site will vary depending on the number of voice channels it can support and the actual channels that are assigned by the RF engineering group. This value can range anywhere from one T-span to three T-spans.

Finally include any POTS lines purchased by the company as part of the operation of the network. Put these data into tabular form and include a column specifying the number of facilities to be purchased or expanded on a monthly basis and the number to be decommissioned owing to low usage or a change in the network configuration. See Table 9.7 for an example layout of these data. This study, once completed, will provide an excellent input to the switch port study discussed in Sec. 9.3.2.

Summary of steps needed to conduct the network interconnect growth study

1. Collect the following data from the various departments in your company.

- Current number of voice circuits assigned in the switch for both land trunks (IMTs and PSTN type) and cell site trunks.

- Current number of POTS owned by the company for use in operating the system. These facilities do not normally affect the switch port capacity since they are not interconnected to the switch.

- Projected number of coverage cell sites and their associated voice channel count to be added to the system over the specified time period from the RF engineering group.

- Projected number of capacity cell sites and their associated voice channel count to be added to the system over the specified time period from the network engineering group.

TABLE 9.7   System Data for Market X for the Network Interconnect (Voice) Growth Study, March 1995

Current number of land (PSTN) switch ports available	2000
Current number of PSTN (LEC) trunks	1000 (approximately 42 T-spans)
Current number of PSTN (IMT) trunks	500 (approximately 21 T-spans)
Current number of cell or RF switch ports available	1500
Current number of cell site trunks	900 (40 T-spans)
Current number of POTS lines	20

- Projected number of PSTN and IMT trunks to be added and decommissioned in the network from the network engineering group. These figures will depend upon the current traffic loads in the network as well as future projected traffic growth and system configuration changes (adding new switches, nodes, etc.).

2. Assemble the above data in tabular form with the system facilities growth and reduction totals shown broken out on a monthly basis.

3. Compare the monthly facilities totals to the current switch port capacities. If there is a shortage in the number of ports needed in any one month, either order more ports for the switch or, if necessary, order a new switch for expanding the network. In a multiple-switch environment it is necessary to move (reassign) cell sites to other switches for purposes of balancing the port assignments and processor load across the network.

RF and network engineering cell site projections. A total of 15 coverage sites and 8 capacity sites will be added to the system over the next 9 months according to the schedule in Table 9.8. Each site will be brought into initial service with one T-span.

Network engineering's system landline (PSTN) facility projections. A total of 240 trunks will be added to the system according to the schedule in Table 9.9.

TABLE 9.8 System Cell Site Projection

Month	Coverage sites	Capacity sites	Month	Coverage sites	Capacity sites
Apr.	2	1	Sep.	1	1
May	2	2	Oct.	2	
Jun.	1	2	Nov.	2	
Jul.	2	1	Dec.		
Aug.	3	1			

TABLE 9.9 System Network Interconnect Facilities Analysis

RF ports required	972	1068	1140	1212	1308	1356	1404	1452	1452
RF ports available	1500	1500	1500	1500	1500	1500	1500	1500	1500
Total ports remaining	528	432	360	288	192	144	96	48	48
Land ports required	1548	1620	1668	1692	1692	1740	1740	1740	1740
Land ports available	2000	2000	2000	2000	2000	2000	2000	2000	2000
Total ports remaining	452	380	332	308	308	260	260	260	260

From Table 9.8 there is sufficient capacity in the system to accommodate the growth in the amount of network facilities growth from a switch port standpoint. Other system considerations must be taken into account when analyzing the interconnect facilities. For instance, is there enough external hardware to support this network growth? DACs and patch panels may have to be ordered as well. The data presented in this study need to be discussed with your finance and/or revenue assurance department for reconciliation of the bills associated with these facilities. The traffic engineer working on the system should be working closely with the finance department in attempting to find ways to route the network traffic over the least expensive facilities. Perhaps by consolidating portions of the network traffic into larger volumes a better rate can be obtained from the LEC? These and other design and cost issues should be discussed among these departments. As a final note, these data are important for the operations personnel to perform their own audit of the network interconnect facilities. They are a great source to rely upon since they work with this equipment on a daily basis.

9.7   Network Interconnect Growth Study (Data)

The network interconnect growth study (data) is similar to the above voice circuit study with the exception that this review pertains only to the facilities used to transmit network-related data for use in call delivery, subscriber validation, etc. The actual data for this study should be obtained from the network engineering group and should include such data as the current number of SS7 data links active in the system, the number of links expected to be added in the system due to network growth and reconfiguration, etc. I suggest that these facilities be kept separate since they are usually ordered and tracked in a different manner than the voice circuits mentioned above. When actually conducting this study, review all the planned projects for the network to determine if new data links will be required, the number of links and hardware needed, and the time frame for delivery of these items. All this information will be necessary to complete the data interconnect study.

9.8   Budgeting

The budgeting aspects to the plan should account for all capital and expense items associated with the network and RF growth plan and should identify the total cash requirements needed for the growth plan period. The actual format for the budgeting issues associated with the plan should correlate with the format used for the technical budget submitted for the time frame this report covers. The budgeting aspects to be included are all new capital and expense requirements, which need to be compared against the current budget already submitted and variances identified.

9.9 Final Report

The final report for the network and RF growth plan is the part of the project where all the efforts put forth to date are combined into a uniform document. Many methods and formats can be used for putting together a growth plan report. A sample network and RF growth plan report is included here to be used as guide for putting your own report together. When crafting the final network and RF growth report it is exceptionally important to remember who is your target audience. The report itself will be used by both upper management and the engineering department to conduct the actual planned network growth.

9.9.1   Example of a network and RF growth plan

The following is an example of a network and RF growth plan that you can use as a boiler plate for putting together your own report.

Growth Study for System X

1.0 Executive summary   The attached network and RF growth study is a continuation of a series of quarterly reports which define the growth requirements for the system. The results from the analysis indicate that for the next year a total of 75 macro cells are required to support the expected expansion of the network. In addition to the cell site expansion requirements a new switch is required and is proposed to be colocated with the existing MTSO location 3.

2.0 Introduction   This study was performed as part of the ongoing effort to determine the network and RF system requirements necessary to support the overall subscriber growth projections, performance improvements, and marketing-driven initiatives.

This report is the latest in a series of quarterly update reports which define the capital and expense requirements of system X. The time frame utilized for this report spans the 1996 to 1998 time frame broken down on a quarterly basis.
The particular components analyzed in the study include cell sites, mobile switching centers (MSC), facility requirements, and all the other auxiliary systems which comprise the system X market. The network interconnect requirements were included in this study. This document is structured in a format that is representative of the steps or building blocks in which it was developed.

3.0 **Design criteria**

Customer Growth Rate	45%	1996–1997
	40%	1997–1998
Baseline Date	June 1995	
Traffic		
Cell Site	Average of 10 busiest days per sector/month	
CPU	Average of 10 busiest days/month	
Subscribers	75,000	
Cell Sites	170	
Radios	5810	
	Switches	2
Dilution	5%	1996–1997
	10%	1997–1998
Max Radios/Sector	19	($N = 7$)

Note:

1. The cells identified in the construction status chart, projected to be operational within the next 6 months, were used as part of the assumptions.
2. Coverage requirements sites identified by marketing and system performance were also included in the analysis.
3. Usage and subscriber dilutions throughout the network would remain in the same proportion for the entire study period.

4.0 **Cell site analysis**

4.1 **Analysis** The growth projections identified in Sec. 2.0 were applied to the baseline sector data and used to determine the radio additions required for each quarter during the study period. Once the radio limit (19) was reached on a particular sector, a new site (search area) would be generated to offload the overloaded sector and support future growth. In addition to the trigger site offloading, all adjacent cells that would be affected by the new site were analyzed and deloaded. After the capacity sites were identified the strategic coverage sites were added and their impact to the adjacent cell was factored in. The radio requirements for each of the sites were then determined.

4.2 **Summary of requirements** As of June 30, 1996, the system consisted of 170 cell sites. There are 21 additional sites planned for the next 6 months. To accommodate the projected growth in usage an additional 34 capacity sites would be required over the next year. The portable coverage requirements for 1997 trigger an additional 20 coverage sites. The results of the analysis included (1) the quantity and locations of new sites, and (2) the quantity of radios per sector and total required for each quarter during the study period. The associated chart with the various breakdowns is included for reference.

System X:

	4Q95	1Q96	2Q96	3Q96	4Q96	1Q97	2Q97	3Q97	4Q97
Area 1	1	3	3	5	7	4	3	2	3
Area 2	1	4	4	5	3	2	2	3	1
Area 3	2	5	4	2	4	2	1	5	1
Area 4	2	3	1	3	2	4	2	1	5

5.0 Mobile switching center analysis

5.1 Baseline assumptions   A set of baseline assumptions were necessary to begin the mobile switching center (MSC) analysis and are as follows:

1. The traffic distribution throughout the system would be based on the distribution resulting from the cell site analysis.
2. Continue to add mods to existing switches until exhaust of processors is reached.
3. The present subscriber database limit of 75,000 would be maintained with future generic releases.
4. The busy hour call attempt (BHCA) limit of XXXX would be maintained with future generic releases.
5. The use of transcoders would be used where necessary to reduce facilities costs.
6. The switch growth will be driven off of the CPU load, subscriber storage capacity, and switch port capacity. Wherever possible any particular switch reaches an upper limit for any of their metrics. Then offloading or further growth for the node or switch must be planned for some 6 months in advance to prevent any problems in the network.

5.2 Analysis   The analysis of the network requirements is included in the following subsections.

5.2.1 Switch CPU load analysis

5.2.2 Switch port capacity analysis   The results of cell site analysis in Sec. 4.0 were provided on a spreadsheet that was manipulated to accommodate planning of all RF derived network requirements, which include quantity of cell site T1's and port requirements. The usage projections were then applied to the public switched telephone network (PSTN), tandem, and cellular networking trunk groups to determine the T1 facility requirements for "network interconnect." The cell site backhaul and network interconnect requirements were combined into a summary of the T1 facility for the study to evaluate the physical limits of the switch. The growth projections were then applied to the traffic-sensitive network elements to determine usage-sensitive trigger requirements.

5.2.3 Switch subscriber storage analysis   The subscriber growth projections were then reviewed to determine the point at which the subscriber database limit would be reached and trigger a new switch. Once a limit was reached on a particular network element and growth addition, a new switch or reconfiguration would be required. In the event that a reconfiguration was required, the T1's and associated traffic would be adjusted to reflect the new arrangement. The data were then trended according to projected growth for the remaining portion of the study period.

6.0 Adjunct requirements   The adjunct system requirements were also analyzed for this growth plan and found to not need any additional dimensioning for the study period.

7.0 **Network capital requirements** The capital requirements necessary to support the network and RF growth plan depicted in this study are summarized in the table below:

	4Q95	1Q97	2Q97	3Q97	Total
MSC (new)	$0	$0	$7,532	$0	$7,532
MSC (growth/mod)	$55.0	$143	$267	$140.0	$605
Buildings	$0	$120	$750	$0	$870
Cell site (new)	$1,269.0	$7,321.0	$1,030	$5,219	$8,250
Cell site (growth/mod)	$660.0	$740.0	$960.0	$420.0	$2,780
Network facility	$225.0	$128	$1,415	$1,765	$3,533
Department capital	$975	$925	$950	$950	$3,800
Total	$3,184	$9,379	$12,924	$8,494	$33,981

9.10 Presentation

When presenting the material to upper management and your fellow engineers it might be beneficial to craft two versions of the presentation. One presentation will be given to the technical departments as a whole. The other is meant for upper management to view. Both presentations should involve a combination of visual aids and handouts. The visual aids should consist of several free-standing charts depicting the current and future network configuration. Additional visual aids include the use of overhead slides describing the key attributes of the plan. The handouts distributed should reflect the exact same information that is shown in the overhead projections.

The engineering presentation should take about 2 to 4 hours to present, depending on whether it is a quarterly update or a yearly plan. The material presented should be of sufficient detail to ensure all the departments within the technical organization understand the general implications of the information in the report.

It is recommended that a member of each department, usually a member of the growth plan, provide a description of their group's projects planned over the report period. The discussion of the various subplans should also include a projected start and end date for each of the topics discussed.

The upper management report is very similar in nature to the presentation that is given to the technical groups of the company. The primary difference in the presentations is the time frame and emphasis of material. The upper management presentation needs to last about 1 hour as the extreme upper limit. Visual aids are the proper method to use for this discussion. It is recommended that the technical aspects take on a very high level approach and not focus on details, unless instructed to do so. The presentation should include a few charts showing the growth trend of the network, critical triggers for the network, new cell sites expected, and cash requirements. The props used for the discussion with technical departments should also be presented as background material.

References

1. AT&T, *Engineering and Operations in the Bell System,* 2d ed., AT&T Bell Laboratories, Murray Hill, N.J., 1983.
2. DeGarmo, Canada, Sullivan, *Engineering Economy,* 6th ed., Macmillan, New York, 1979.
3. Kaufman, M., and A. H. Seidman, *Handbook of Electronics Calculations,* 2d ed., McGraw-Hill, New York, 1988.
4. Keller, Warrack, Bartel, *Statistics for Management and Economics: A Systematic Approach,* Wadsworth, Belmont, Calif., 1987.
5. Lee, C. Y., *Mobile Cellular Telecommunications Systems,* McGraw-Hill, New York, 1989.
6. Lindenburg, *Engineering Economics Analysis,* Professional Publications, Inc., Belmont, Calif., 1993.
7. MacDonald, "The Cellular Concept," *Bell System Technical Journal,* vol. 58, no. 1, 1979.

10

Organization and Training

The organization of the technical departments and their training will directly influence the degree of success or failure the communications company experiences. This chapter addresses two very important aspects of any organization. The first focuses on the organization structure with specific responsibilities delineated. The second focuses on the training requirements needed for the engineering departments.

10.1 Technical Organization Structure

The technical organization structure utilized by your company should be driven by functional requirements instead of personalities. Personalities, however, often define the organization's structure in that the organization is arranged by who is in the group and not who should be doing the work.

The technical organization can take on a centralized role or a distributed role for the company. It is recommended that a blended approach be utilized for the company's technical structure. The concept of using a centralized versus a decentralized approach is largely dependent upon the company's culture. The centralized approach has the advantage of potentially achieving economies of scale. The economies of scale are achieved through the elimination of redundant functions in each of the markets or areas. An example of a centralized function is new technology research for the company, which does not have to be accomplished with every department or division within a company. The idea of having one group lead the effort will ensure uniformity, accountability, and the probability that the direction picked is coordinated among the various organizations in the company.

The centralized approach, however, has the disadvantage of being defocused on the market requirements. This can come about through not having any local knowledge of the technical configuration for the network. An example of defocusing can occur over a simple matter of switch port assignments or roads a cell site actually covers.

The centralized approach is also susceptible to overwhelming the engineers in their tasks and areas of specific responsibilities. It must be used with caution when supporting multiple technologies, which involves having the same workforce support a varied range of technologies like CDPD, PCS, and technology trials all in addition to supporting cellular technology.

The decentralized approach has the advantage of being more market-sensitive and flexible than the centralized approach. An example of its flexibility involves the continuous configuration of the network based on handoff and call-origination traffic. The idea of the decentralized approach is that the decisions that will affect the market are brought as close to the customer as possible.

The disadvantage with the decentralized approach is the amount of redundant work that is performed. The decentralized approach lends itself to localized procedures that foster inefficiencies and a lack of knowledge transport, which often leads to the problem or situation being repeated in another market when some simple communication could convey how it could possibly be avoided. The decentralized approach also does not lend itself to any engineering practice procedures, which is essential in the rapidly changing world of wireless communications. However, a straight centralized or decentralized approach is not necessarily best. The organization structure for the technical organization should take on a blended approach to centralization and decentralization.

10.2 Centralized Technical Organization

A centralized technical organization is shown in Fig. 10.1. The organization proposed can take on a few variants in the approach. Specifically the variants pertain to whether the company has several divisions or does not have any other divisions. If a company has another division, say, a west coast operation and an east coast operation, each would be considered a different market and would have its own technical directorate. The centralized organization proposed in Fig. 10.1 involves a span of control that can range from five to possibly seven or eight. Exceeding a span of control beyond seven or eight for any organization becomes very difficult to manage for any prolonged period of time.

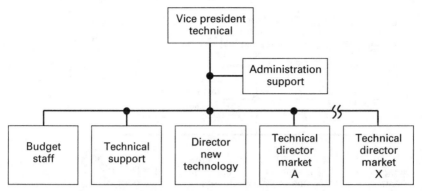

Figure 10.1 Centralized technical organization.

The key roles listed in the proposed technical organization involve a multitude of disciplines which comprise a well-rounded organization. The disciplines needed for the organization shown are

1. Budgetary
2. Technical support
3. New technology
4. Technical directorates

The budgetary roles are straightforward and involve tracking, variance resolution, and forecasting the capital and expense budgets for each of the various groups under this organization. The budgetary group in the organization plays a vital role in helping secure the necessary funding for various projects.

The technical support group's role is to establish, track, and report on the individual network's performance. Their role is to be a central clearinghouse for technical performance requirements leading to a standard set of criteria for all to use. This group also would ensure a best practices approach is accomplished among all the markets, allowing for good ideas and procedures to be shared.

The individual technical directorate would be responsible for the market's technical performance. The specific groups the directorate would be responsible for are shown in Fig. 10.2. The particular roles for each group under the technical directorate involve budgets, engineering, real estate and implementation, and operations. The budget staff listed in the chart are responsible for tracking, variance reporting, forecasting, and purchase order handling for the entire technical organization. The director for engineering is responsible for the design of the network and the technical performance aspects. The

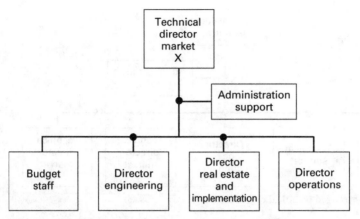

Figure 10.2 Centralized technical organization by market.

director for real estate and implementation is responsible for the acquisition, leases, civil work, and construction of the various projects put forth by engineering. The operations director is responsible for ensuring that the equipment installed in the network is maintained and operating at its peak performance possible.

The proposed flow of work or information regarding an organization is shown in Fig. 10.3. It is not all-encompassing because of equipment problems found by operations and fed back into the various technical departments for problem resolution and/or redesign.

10.3 Decentralized Organization

The decentralized organization is shown in Fig. 10.4. The organization proposed can take on a few variants to the structure listed.

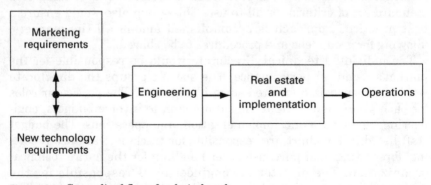

Figure 10.3 Generalized flow of technical work.

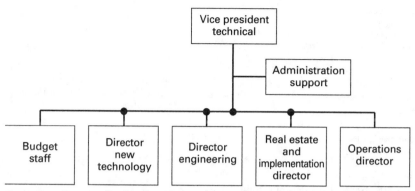

Figure 10.4 Decentralized technical organization chart.

Specifically the variants pertain to whether the company has a centralized budget department for all the organizations. The decentralized organization proposed in Fig. 10.4 involves a proposed span of control of five or rather six if you include the administration support. As with the centralized organization, exceeding a span of control beyond seven or eight becomes very difficult to manage for any prolonged period of time.

The key roles listed in the proposed technical organization involve budgetary, new technology, engineering, real estate and implementation, and operations directorates. The budget staff is responsible for the capital and expense tracking, variance reporting, forecasting, and purchase order handling for the entire technical organization. The director for engineering is responsible for the design of the network and the technical performance aspects. The director of real estate and implementation is responsible for the acquisition, leases, civil work, and construction of the various projects put forth by engineering. The operations director is responsible for ensuring that the equipment installed in the network is maintained and operating at its peak performance.

The proposed flow of work or information regarding an organization is shown in Fig. 10.5. It is the same flow shown for the centralized organization.

10.4 Engineering Organization

The engineering organization, regardless of whether it is part of a centralized or decentralized hierarchy, needs to ensure the network is operating at the desired performance criteria. The engineering department in the organization needs to take the leadership role for

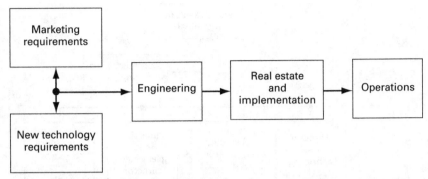

Figure 10.5 Generalized flow of technical work.

defining the overall direction of the technical community. The engineering organization and responsibilities should be driven by market requirements.

The engineering department is responsible for the planning and design of the communication system from an RF and network perspective. The proposed organization for a typical communication engineering directorate is shown in Fig. 10.6. It represents an organization that is needed for a large metropolitan system. Many variants to the proposed engineering organization structure are possible and do exist currently. When looking at a possible variant to this proposed organization the individual responsibilities of the groups need to be properly defined.

The network systems group is responsible for the architectural engineering of the network growth as well as evaluating new network designs and performing network troubleshooting. The network system group plans the switch dimensioning, module growth, and the proposed location when a new switch is needed for the network. The group also is responsible for value-added systems that are adjuncts to the network and require interfacing to the actual network. Another area of direct responsibility for this group ties into the recorded announcements of assignments associated with any network. Additionally the group is responsible for all product development design aspects for the network that are derived by marketing.

The facilities management group is responsible for all aspects associated with facilities and utilities management. The group is responsible for ensuring the trunk designs for the network are adequate and properly dimensioned for growth. Another role the group is involved with is ordering the actual facilities for the network and ensuring they arrive at the predetermined time for the project. The group is also responsible for dimensioning the DACs in the network and interfacing all the issues associated with the IXEs.

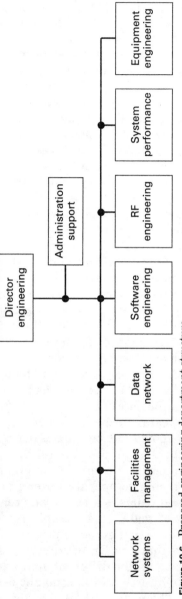

Figure 10.6 Proposed engineering department structure.

The data network portion of the engineering group is responsible for protocol issues associated with the network. Some of the protocols utilized in the network are IS-41, ISUP, SS7, and MTUP, to mention a few. This group plays an important and pivotal role for managing the out-of-market interfaces of the network.

The software engineering group is responsible for all the software loads put into the network and its impact. The group is responsible for leading any software FOA for the switches and cell sites. The group oversees all major feature introductions into the network. The software engineering group also produces all the data translation information that is used for the switches and cell sites. Another aspect the group is responsible for involves validation of the billing system whenever there is a software load change for the cell sites or switch. This group also performs system software audits to ensure that the software and translations are consistent across multiple nodes in the network and removes all unwanted data.

The RF engineering group is responsible for the macro and micro cell planning and design efforts. The group also is responsible for the frequency management of the network, intersystem coordination, and FCC and FAA compliance for new sites.

The system performance group is responsible for the overall performance of the existing RF network of the system. It is also responsible for ensuring that all the new cell sites in the network perform at a specified quality control standard. The group must also interface with customer care for problem ticket resolution.

Equipment engineering is responsible for the physical cell site and switch equipment designs. They are responsible for ensuring configuration management in the network.

Considering many years of experience I recommend that the engineering organization should not be broken down into a more decentralized function than that proposed here. Specifically the current trend is to operate in localized teams for an organization. The major disadvantage with trying to operate an engineering department through use of localized teams is a loss of any configuration management and economies of scale. It is strongly recommended that the engineering directorate be configured as proposed.

It is recommended that the RF engineering and system performance groups operate under different managers. The rationale behind this effort ties directly in the constant demand for resources for the build program of any network. It is essential for the short- and long-term health of the network that these two groups remain independent of each other so that a check and balance system is put into the network design. With the large amount of capital and expense dollars expended on a yearly basis it is imperative that some form of such a system be instilled in the organization.

10.5 Technical Training

Technical training for any organization is a difficult task since there is never enough time to ensure proper training of personnel takes place. This section has a list of proposed general topics that should be utilized by the various engineering departments to ensure that there is a minimum level of formal training for all the groups.

It is strongly recommended that every new and existing employee be assigned a mentor for guidance. With the demand for immediate results for any issue, the use of a mentor program as an augmentation to formal training will expedite the training of personnel. The selection of qualified mentor individuals needs to be taken with extreme caution to ensure that bad design practices are not propagated to the new members of the organization.

The training proposed for the engineering organization is defined in three sections, management, RF, and network. The rationale behind not selecting individual training programs for each of the individual organizations listed above is multifaceted. First is the need to have a certain degree of cross training between the major groups in engineering. Second is the focuses on the potential variances and individual market requirements for every organization.

The training that takes place for all the members of a technical organization should involve around 80 hours of training per year. The 2-week training commitment is necessary for any technical organization to remain current and ensure its personnel are adequately trained. It is strongly suggested that you employ a training deconfliction system to ensure that everyone receives training in the year and at the same time reduces the level of personnel outages caused by training.

The proposed training deconflicting schedule is to list all the training expected to take place for the employee and the department as a whole for the entire year. It is recommended that the schedule deconfliction have an initial granularity of monthly as a minimum. Regarding the training deconfliction schedule, it is also important to list along with the training the personnel vacation plans, conferences, and major project milestones.

The following is a brief listing of suggested technical material that needs to be taken by various major subgroups within engineering. The list provided, however, does not include the courses required by your company like orientation and various diversity classes, to men-

Training Deconfliction Report

Month	Jan. 96	Feb. 96	etc.
Employee 1			
Employee 2			

tion a few. It also does not define the number of classroom hours needed for each topic, or the repeat interval for technical material. It is important to remember that certain courses should be required to be retaken at regular intervals to ensure that a minimum level of competency is maintained at all times.

Management

- Basic Cellular Communications
- Budget Training
- Cellular Call Process
- Digital Radio Design
- Disaster Recovery
- Fraud Management
- General Management
- Interconnect
- Interview Techniques
- IS41
- MFJ/Descent Decree
- Network Design

- Network Fundamentals
- Operations and Maintenance
- PCS
- Presentation Techniques
- Project Management
- Real Estate Acquisition
- RF Design
- SS7
- Statistics Theory
- Switch Architecture
- Tariffs
- Traffic Theory

Network engineering

- AIN
- ATM
- Basic Cellular Communications
- Call Processing Algorithms
- CDPD
- Cellular Call Processing
- DACs
- Data and Voice Transport
- Disaster Recovery
- Equipment Grounding

- Network Fundamentals
- Network Maintenance
- Network Services
- Numbering Plans
- PCM and ADPCM
- Performance Troubleshooting
- Presentation Techniques
- Project Management
- RF Design
- SONET

Fiber Optics	SS7
Inter- and Intra-LATA	Statistics Theory
Interconnect	Switch Architecture
Interview Techniques	Switch Maintenance
ISDN	Tariffs
IS41	Traffic Theory
LAN/WAN Topology	Translations Switch
MFJ/Descent Decree	Translations Cell Site
Network Architecture	Voice Mail
Network Design	WPBX

RF engineering

Antenna Theory	Microwave System Design
Basic Cellular Communications	Network Architecture
Call Processing Algorithms	Network Design
CDMA	PCS
CDPD	Performance Troubleshooting
Cell Site Grounding	Presentation Techniques
Cell Site Installation	Project Management
Cell Site Maintenance	Real Estate Acquisition
Cellular Call Processing	Rerad Design
Digital Radio Design	RF Design
Disaster Recovery	Statistics Theory
EMF Compliance	Switch Architecture
Frequency Planning	TDMA
Interconnect	Traffic Theory
Interview Techniques	WPBX
Microcell Design	

Every employee, new and old, for a department needs to have a training program. The training program is more of a development program with the desired intention of helping improve the skill sets of the individual. In addition to improving the skill sets for the individual the department as a whole will be improved through higher efficiency levels obtained with the new skill sets.

When crafting and authorizing any training program it is essential that the subject matter is relevant to their training program and job function. However, when a new employee arrives it is imperative that they be given some guidance to help ease the transition period. A recommended format for establishing a new hire training program is listed below. The details for each section will need to be tailored for the particular job function, but the general gist of the program can be easily extracted here.

To: Name of employee
From: Manager
Date:
Subject: Training Program for Position X

Objective:
  Define here what the stated objective is for the training program and its duration.

Description of job:
  This is where a description of the actual job would be specified.

Reports:
  Description of the various reports needed to be generated by the individual and when they are due and to whom.

Reading material:
  This is a brief description of the suggested reading material the person should have in their personal technical library. It is imperative in this section that the source locations for the various documents be spelled out.

Training program:
  This section involves putting together a training program for the individual. The mentor assigned to the person also needs to be defined.

Assigned project:
  This section lists the particular projects assigned the individual and their deliverables, with a time line.

Manager signature: _____

Employee signature: _____

The rationale behind signing the training document is to ensure that the scope and responsibilities for the employee are clearly understood. The training document should be an integral part of the employee's yearly goals and objectives.

References

1. Aidarous, Plevyak, *Telecommunications Network Management into the 21st Century,* IEEE Press, New York, 1993.
2. AT&T, *Engineering and Operations in the Bell System,* 2d ed., AT&T Bell Laboratories, Murray Hill, N.J., 1983.
3. Brewster, *Telecommunications Technology,* Wiley, New York, 1986.

4. Chorafas, *Telephony: Today and Tomorrow,* Prentice-Hall, Englewood Cliffs, N.J., 1984.
5. DeGarmo, Canada, Sullivan, *Engineering Economy,* 6th ed., Macmillan, New York, 1979.
6. Dixon, R., *Spread Spectrum Systems,* 2d ed., Wiley, New York, 1984.
7. Hess, *Land-Mobile Radio System Engineering,* Artech House, Norwood, Mass., 1993.
8. Jakes, W. C., *Microwave Mobile Communications,* IEEE Press, New York, 1974.
9. Keiser, Strange, *Digital Telephony and Network Integration,* Van Nostrand Reinhold, Princeton, N.J., 1985.
10. Keller, Warrack, Bartel, *Statistics for Management and Economics: A Systematic Approach,* Wadsworth, Belmont, Calif., 1987.
11. Lee, C. Y., *Mobile Cellular Telecommunications Systems,* McGraw-Hill, New York, 1989.
12. Lindenburg, *Engineering Economics Analysis,* Professional Publications, Inc., Belmont, Calif., 1993.
13. MacDonald, "The Cellular Concept," *Bell System Technical Journal,* vol. 58, no. 1, 1979.

Index

ABOUT THE AUTHORS

CLINT SMITH, P.E., is currently the vice president of Communication Consulting Services (CCS) and is a registered professional engineer. He has over 14 years of radio engineering and management experience with 9 years directly in cellular communications. Prior to CCS, Mr. Smith was the director of engineering for NYNEX Mobile in New York where he was responsible for RF, network, system performance, software, and data transport technologies. Clint holds a master's in business administration from Fairleigh Dickinson University and a bachelor of engineering degree from Stevens Institute of Technology.

CURT GERVELIS is a senior staff engineer for CCS with over 9 years experience in the cellular and wireless telecommunications field. His prior experience includes 5 years working for Cellular One of New York as a systems engineering manager, project engineer, and senior systems engineer and 3 years in Motorola's Cellular Infrastructure Division as a senior systems engineer. Curt holds his master of science in electrical computer engineering from the University of Cincinnati and his bachelor of science in physics and mathematics from Mount Union College.